Data Governance

Data Governance

How to Design, Deploy, and Sustain an Effective Data Governance Program

Second Edition

John Ladley

ELSEVIER

ACADEMIC PRESS

An imprint of Elsevier

Academic Press is an imprint of Elsevier
125 London Wall, London EC2Y 5AS, United Kingdom
525 B Street, Suite 1650, San Diego, CA 92101, United States
50 Hampshire Street, 5th Floor, Cambridge, MA 02139, United States
The Boulevard, Langford Lane, Kidlington, Oxford OX5 1GB, United Kingdom

Library of Congress Cataloging-in-Publication Data
A catalog record for this book is available from the Library of Congress

British Library Cataloguing-in-Publication Data
A catalogue record for this book is available from the British Library

ISBN: 978-0-12-815831-9

For information on all Academic Press publications
visit our website at https://www.elsevier.com/books-and-journals

Publisher: Mara Conner
Acquisition Editor: Mara Conner
Editorial Project Manager: Peter Llewellyn
Production Project Manager: Punithavathy Govindaradjane
Cover Designer: Greg Harris

Typeset by SPi Global, India

Working together to grow libraries in developing countries

www.elsevier.com • www.bookaid.org

To Pam
...more today than yesterday, but not as much as tomorrow
Tum tee tum tum, tum tee tummmmmm ......

Contents

Forewords
Today's Organizations are "Unfit for Data"

For some time, I have been frustrated by the slow pace in the data space. There are many "points of light," but so far not enough really big wins. Some of my frustration is aimed at data governance. Practically every organization has a data governance program. Yet most data is of poor quality, companies have way too many disparate copies of the same stuff, and security and privacy breeches are all too common.

The more I look, the more I find organizations, despite well-meaning programs, that are not defining the issues properly, don't have the right people in the right roles, and are attacking the issues incorrectly! Some years ago, I started pointing out that organizations are "unfit for data."[1] It is no less true today. But it is more important!

The following vignette summarizes one fundamental error made by too many proponents of data governance:

Data Governance Proponent (DGP): "First, we need to pick the best tool."

Me: "We should think about this carefully. Here is a little model for developing data programs that I've found helpful:

$$\text{Strategy} \rightarrow \text{Organization} \rightarrow \text{Process} \rightarrow \text{Technology}$$

"So let's first make sure we're clear about what we're trying to achieve. Then let's get the organization right. After that we can discuss the technology."

DGP: "Yes, Tom. I see what you mean. That is a great idea."

DGP (30 min later): "Wow, this is really hard. And now that I think about it, that strategy and org stuff is for the business side. We're responsible for the technology. So, let's go ahead and pick a tool."

It is almost impossible to recover from such thinking. Even worse, I can go on and on with such vignettes.

With data governance's mixed record in mind, earlier this year, I asked John Ladley, "Is it time for a fundamental re-think of data governance?"

His reply shocked me, "No, Tom," he said. "It is not time for a fundamental re-think. It is time for a fundamental THINK. It is rarely thought out in the first place!"

I'm glad John is on the case.

First and foremost is the question, "What exactly is data governance?"

I've made the statement that, from the point of view of a software vendor, "data governance is the stuff you have to do so our system doesn't fail." I'm being a touch sarcastic, but many have agreed that it is a good operational definition.

Others treat data governance as the means to meet regulatory requirements, to develop a common vocabulary, and so forth. Net, net, data governance is a confused mess.

John sorts it all out. Suppose you are a senior corporate executive. How would you know if your company's data program(s) meets the needs of the company? And if not, what should you do about it? These are the questions, the only questions, that data governance must address.

---

[1] See, for example, "Are you Ready for a Chief Data Officer?" https://hbr.org/2013/10/are-you-ready-for-a-chief-data-officer

At its heart, governance is about control at the corporate level. For most, data governance represents a new business capability, unlike anything they have in place now. Growing that capability and building it into the fabric of the company will be a real challenge and take a lot of time. It will also take courage.

John gets these fundamental points right. He goes on to tell you who must be involved and how to do the work. One key point: Direct, personal work by the organization's most senior leaders is essential! I've heard many middle managers blithely proclaim that "we have executive support." That is simply not sufficient.

I hope readers see both risk and opportunity. The most important risk is that your data governance program will continue to muddle along. The opportunity is to be first in your sector to get data governance right, thereby creating an advantage.

Go for it!

Thomas C. Redman
the Data Doc, Rumson, NJ, United States

It takes a special kind of person to really LIKE data governance. After all, this discipline exists at the epicenter of data-related conflict. Day after day, we see how seemingly small actions and decisions create data-related problems that ripple out through an organization, creating bigger problems in reports and other information products, which create even bigger problems in the form of bad decisions, inefficiencies, ineffective practices, noncompliance with laws and regulations, and even security breaches. We stand our ground, watching these problems as they are created, as they grow, and as they impact our organizations' abilities to meet their missions. We engage the people around us, trying to educate them about how to avoid creating those problems, how to find them, and how to fix them. We work with C-suite executives, individual data workers, and everyone in between, preaching the same message over and over: *"You don't have to live with the consequences of bad data. Let us show you a different way."*

But, frankly, most people don't want to hear it.

Most don't love data for its own sake, just for what it does for them. Most people hear the word "governance" and have a negative—even visceral—reaction. Their rational mind might be promoting the idea that "Big G" governance mechanisms (policies, mandates, standards, control objectives, and other types of rules) are necessary. They might rationally agree that "little g" governance mechanisms (controls) are essential. Still, their nonrational, emotional, primal brains will be reacting predictably to any constraints, calling for the listener to fight, flee, or play opossum.

So imagine how delighted I was to meet John Ladley, someone who addresses the human aspects of governance adoption from an anthropologist's perspective, its strategic aspects from an executive's perspective, and its operational aspects from a practitioner's perspective.

I think I heard John laugh before I ever heard him speak. It was at a conference, and someone had just said, "No, they don't want the responsibility [of data governance], but they don't want anyone else to have it either!" John's laugh was contagious, and his face lit up at this example of human nature. He followed up with some words of wisdom regarding organizational change management, and we got into an extended discussion about details concerning some information management strategy that I don't remember now. Later, I discovered that his thought leadership came from a vendor-neutral perspective and a strong sense of intellectual integrity. John has been a part of my personal "Kitchen Cabinet"—as well as a personal friend—ever since.

The funny thing about data governance is that it is both old and new. When I was working in publishing in the 1980s, we didn't have automated workflows. We had hundreds of chunks of information that had to go through multiple iterations and alterations before finally being compiled into a magazine with a specific number of pages. If our content chunks weren't well governed, we couldn't deliver our product. Our mailing lists and other structured data had to be well governed, or we couldn't operate. Oh yes—ask anyone who was working in publishing (or working with mainframes) 30 years ago, and they'll tell you: Data governance was just a part of doing your job back then.

It was the rapid explosion of IT that changed things. In the rush to move to client-server, web-based, and other game-changing technologies, many organizations lost both "Big G" and "little g" capabilities. The focus of IT became the "T" (technology). In rapidly evolving organizations, it seemed like no one group was responsible for the "I" (information). Things got messy, and then they got messier. Somehow, the problem got labeled as poor collaboration between "Business" and "IT." It took the Sarbanes–Oxley Act of 2002 and an ever-increasing number of data breaches to direct attention back to data and the need to properly govern it.

While John and I (and others on the same conference circuit) had many enjoyable discussions about this "new emerging field" of data governance, I have to confess that I really didn't get why John was also devoting his time to writing his previous book on EIM (enterprise information management). After all, I said, the fields of data management and document/content management are pretty well defined. Do we really need this new acronym? Do we need a new book on the topic?

As it turns out, we did. John brought to that book an important new perspective. This work was not merely instruction for data geeks who loved their little slice of data heaven and were happy to learn about other slices. No, his book also looked at this broad field from the perspective of someone who is used to managing large and important resources for the betterment of an enterprise. This was "Business meets Information Management," with a lot of detail. Yes, it needed to be written. And I was glad John did.

The visits to the Data Governance Institute's website told the story of the ever-growing number of people who were getting engaged in governance. And even though more and more of my consulting time turned to helping organizations with strategies, I wanted to talk about data governance practices. Selfishly, I was glad when John wasn't working on his EIM book any more so we could have DG discussions. In a world where governance has so many focus points, and so many different "flavors," I would ask him, what is universal? What is situational? What is need-to-have, and what is nice-to-have?

John is a man of action, so he often countered my topics with ones about specific activities and action plans: the HOW of data governance.

In the past several years, much has been written about why organizations need data governance, and who should do what, and how to sell the concept to those with the budget to fund projects and programs. Much has also been written from tool vendors' perspectives, and much has been written from a motivational perspective. But not much has been written about the details of WHAT to do, and WHEN, and HOW. The world needed a big detailed instruction manual—one that would be relevant in many situations, for the many "flavors" of data governance.

I'm glad John Ladley has written it.

Gwen Thomas
Founder and President
The Data Governance Institute
www.datagovernance.com

Preface

Welcome to the second edition of this data governance (DG) book. Frankly, data books rarely get a second edition. But DG is not only "hot," it is required. When the first edition came out, DG was a misunderstood capability and few people knew anything about standing up a sustainable DG program. Now, if you examine any major corporate or organization initiative and you will find DG. It most likely is not called governance. Some sort of data oversight is built into projects, or a business area. But it is DG.

Why do a second edition? There are several reasons.

Even though DG is appearing in many forms, it is like volunteer corn in a farmer's fallow field. You see the individual instances appear and provide hope for some corn, but no one will be able to harvest the benefits of a crop. Similarly, most current versions of DG are not sustainable, but the acceptance of the core idea is universal. Therefore, the main reason for the second edition is to continue to educate on how to make DG a sustainable business capability in your organization.

Second, we have learned a lot since the first edition. There is the usual evolutionary maturity that comes from working on DG programs. But data has leapt from the initial whispers of analytics a few years ago to a much-ballyhooed source of wealth and cash flow. CxOs toss about terms like artificial intelligence and data monetization. And most of them have no clue what lies behind success in these areas. I firmly believe leadership requires education. Treating data as an asset is not something to delegate. Building data into products and processes, i.e., creating data capabilities, means thinking about the data aspect of all projects and programs. Data literacy needs to improve at the highest levels of all organizations. I love to quote my friend Dr. Tom Redman—"most companies are unfit for data."

Which segues to the third reason to do another edition. The blind rage continues, unabated.

This is a bit tongue in cheek, but only to a point. Organizations now have a sense that data and information require more than just a few tools to move and cast data about the company. Realizing you need to do something, and then actually doing it, are two different things. I find many organizations are very good at saying, "We are going to do better with data," and they present myriad reasons and justifications for this. But their follow-through is abysmal. And I mean abysmal. Numerous sources show that from 75% to 85% of artificial intelligence and analytics efforts fail to achieve expectations the first time (or two). This includes all of the data science and big data stuff.

But the headlong rush to go out and buy DG and data presentation tools continues. In 2012 on this very page in the first edition I said, "At the time this was being written, vendors were spending gobs of marketing dollars on the value of analytics and 'big data.' Companies are drinking the Kool-Aid™ deeply, but very few reap the anticipated benefits."

And it has not changed. And the keepers of the new business kingdoms, i.e., the data people and their constituents, are dumbfounded at the lack of movement on the part of leadership.

There is realization that something better needs to be done with data. There are many regulatory changes such as CCPA and GDPR. The US federal government has started to implement a sweeping data strategy. Numerous numbers of Chief Data Officers have been hired and fired. But all of this support then gets delegated to the lowest, least empowered levels of organizations.

The answer is the same as it was 2012; start to treat data and information as an asset. Do DG as part of taking care of that asset. Make it an acknowledged business capability.

Even a modicum of discipline will reap benefits. Even with the high failure rates we see the evidence that data management and governance works. Successful efforts exhibit good sponsorship, business alignment, management of organizational issues, and realistic application of technology alongside a flexible roll-out plan. And all are sustained by DG. When you examine an almost identical effort where DG was not applied or was implemented poorly, you see the failures.

For many years excellent work has been done to make DG agile, less invasive, show more immediate benefits, etc. All of the stuff you do with any new methods or solution. As you will discover in the pages ahead, DG is not setting up some processes and policies and enforcing some rules. You need to acknowledge it is a new capability, and at some point, you need to change how some things are being done. DG will not stick unless you embrace doing new business capabilities.

This edition is still for those who need to "do data governance." It is not just for IT or data people. It is for anyone who has to make sure data and information management is happening. To be clear, this is a "how to" book. I tried very hard to eliminate the bromides you can easily hear from a tool vendor or big-name consultant. If you are reading this book, you have heard the platitudes, embraced them, and now want to do something about it instead of talking.

As we go to production with this edition, the deluge of adverts proclaiming the power of data, and algorithms continue. But so do the panicked glances being exchanged between the rank and file data workers who have to make it all happen. "The data is awful," they whisper. "It won't work."

If we continue to treat data as the ugly lubricant of departmental business processes instead of the precious asset it is, we will come nowhere near to fulfilling these forecasts. None of it is possible without significant changes in mindsets. Here are two scenarios you should consider:

- Avoid the issue—A company that budgets for fines and excessive expenses due to incorrect data going to key regulators. Rather than fix the data, they feel it is less work to just pay the fines. OK, I guess, except this is a company whose entire culture and brand is built around quality products and service.
- Address a new business capability—A company that decided to get serious about its information, and the CEO became data literate. He then proceeded to call out every major department head for a period of years when they took the easy way out with their data (i.e., the spreadsheet). They have documented tens of millions of benefits directly to their bottom line in efficiencies and quality-based results.

The term *maturity* is often tossed about in the context of managing information. This book was written with that in mind, but also with another scale—that of learning maturity. My weekend hobby is aviation. I also teach other people how to fly, and I learned a great definition for learning when I became a flight instructor:

Learning occurs when you see a change in behavior as a result of experience.

In other words, just hearing about something is not going to create learning. You need to do it, develop experience, and then look and measure for the change. Frankly, most companies I deal with want a 2-week assessment, a four-week road map, and then they somehow think these artifacts and a few hearty commands from management will work miracles. DG will require some work and some significant behavior changes. This book is written with an eye toward changing behavior and assimilating and managing the work to be done.

The following pages present steps, artifacts, techniques, and insights developed over the past 20 years or so. Some of this material can be incredibly dry, so if I sprinkle in a story or amusing metaphor, it is not because I am overly glib. It is because I really want you to pay attention. *This stuff really matters.* Your organization is going to live or die based on how it deals with data.

The following chapters present a comprehensive view of the work and behaviors required to implement DG. It is presented to allow you to easily configure an approach that works for your situation. I have tried to avoid the impression of a methodology.

Does your organization want to do advanced predictive analytics? You had better know that the data used by the analytics tools is accurate. Do you want to create single sources of truth for reporting, business intelligence, or just getting your customer list nailed down? Then you need to start DG *now*. The longer you wait, the harder the decisions will be as the data explosion continues. This is not a trivial request from someone who likes working with data. This is a business imperative.

You will see that DG can be accomplished by executing a series of steps along with consideration of certain success factors. There are also cultural, personal, and philosophical changes required to truly treat information as an asset. DG is the discipline that encapsulates these changes. One thing that has not changed since the first edition: DG is about control and capability around data. It remains a *long-term commitment to doing business differently*.

John Ladley

Acknowledgments

Since the first edition I have remained fortunate to continue to work with very smart people.

My coworkers and partners in information management and data governance since the first edition are many. Val Torstenson still stuck with me and is still one of the unsung heroes of the data profession. Pam Thomas (in full disclosure, now Mrs. John Ladley) also continued with me until her recent retirement.

Along the way my firm merged with First San Francisco Partners. Kelle O'Neal, Malcom Chisolm, Gregg Loos, Angie Pribor, and the rest of the FSFP crew were a delight to work alongside. It can be risky to have that many smart people in the room. But with a shared purpose it went well, and we did great things. Thanks again to FSFP for allowing redacted versions of some sample artifacts.

For personal reasons, the road to this edition was tortuous and I would like to thank my editor Peter Llewellyn for the patience of a saint.

For the same personal reasons, I also moved away from main line consulting and more into advisory work and research. But along the way, myself, Tom Redman, Danette McGilvray, James Price, and Doug Laney initiated a movement of sorts called DataLeaders to encourage the awareness and expansion of data management (www.dataleaders.org). A lot of their counsel and input appear in this edition.

Dr. Tom Redman's guidance to just tell things directly has been most useful and he kindly wrote the foreword for this edition. Danette's thoroughness, dedication and experience is an invaluable influence and James Price's passion about all things data keeps us all going.

Doug Laney and I have worked together and fed off of each other for a while and I appreciate his insights. Doug was able to crack the door to the executive suite with his book *Infonomics*. We do not agree on everything, but it is a healthy repartee.

I have to thank all the clients who provided the experiences that could be morphed into advice in this book. In addition to those mentioned in the first edition, I would like to especially recognize the achievements of Salt River Project, Vanguard, and Scottrade. I have seen a lot of good work, but those are standouts.

Individuals along the way who asked hard questions and kept me on my toes include Barbara Forth, Diana Aresu, Christy Villa, and Robin Grimwade.

A special shout out to Stacey Clark. I have never seen such single-minded dedication to data governance with such quick returns and visible value as the program he ran. Sadly, his company was acquired, and his program was cancelled. Luckily other organizations are benefitting from Stacey's dedication and understanding of data governance.

On-going thanks to the various people I get to share the label of guru or thought leader. Loretta Mahon-Smith, Gwen Thomas and Dr. Peter Aiken for being excellent sounding boards. Rob Seiner and his noninvasive approach have given many organizations a practical start towards data governance. Rob and I continue to surprise people that Pittsburgh can turn out clever people. Daragh O Brien of Castlebridge gets recognition for sharing terrific insights on data ethics (and a wicked sense of humor) along with colleague Dr. Katherine O'Keefe.

Most importantly, if it weren't for Pam Thomas—my significant other, and total sweetie—this would not have happened. She swore if I ever did another book, she would toss me out. I skirted the issue with a second edition, and she let me get away with it. I am still blessed to have my significant other by my side in this endeavor.

Prologue: An executive overview

Chapter Outline

Companies are unfit for data.
Dr. Tom Redman

The situation

In today's business environment, there is definite awareness that data needs to be managed and governed. Make no mistake. A 21st century organization needs to manage data as an asset. This is the essence of what it means to be "data driven." Many organizations believe they are on the data driven trajectory, and most are really headed in the wrong direction.

The audience of this chapter is an organization's top leadership. To be clear, it does NOT mean the top data roles. It means the folks in the Executive Suite.

As this edition is written there are daily news events where poor data management (DM) has drastically affected organizations. Fraud and security breaches are just the tip of the iceberg. What I am really talking about are the sneaky, slow "boil the frog" type events that organizations slide into. Poor data quality loses customers, item number inaccuracy overstates inventory. Hiring scores of business and data analysts to "do data stuff" is becoming expensive and creates internal fiefdoms. The costs of errors and lost opportunity is documented to be in the $US trillions. That is with a "T." I will show you the exact numbers in an upcoming chapter.

On the positive side, treating your data as an asset offers myriad possibilities. "Each of these capacities represents a discernible, discrete economic benefit which can be monetized, managed and measured. And when any of these go unattended, you're leaving money on the table."[1] Data governance (DG) will oversee the operation and evolution of essential capabilities to manage data assets.

There are risks as well. Besides privacy breaches, errors, reputation, we have entered an age where organizations need to be aware of enormous ramifications of mistreating data: "…we cannot lightly introduce powerful technologies that have the potential to deliver significant benefits to individuals or to society, but equally have the potential to inflict great harms. The complication we face in the information age is that a failure to implement technologies with the appropriate balances in place (e.g. easy to configure security in IoT devices, appropriate governance in analytics planning and execution) has the potential to affect many thousands, if not millions of people, directly or indirectly, before remedies can be put in place."[2]

---

[1]Laney, Douglas, "Infonomics," Gartner-Bibliomotion, 2018.
[2]Katherine O'Keefe, Daragh O Brien, "Ethical Data and Information Management: Concepts, Tools and Methods," Kogan Page Limited, 2018.

Data Governance. https://doi.org/10.1016/B978-0-12-815831-9.00001-1

But often, and I speak directly to organization leadership here, that is where it ends. Data is declared as important, and everyone goes back to work. Organizations experiencing the issues listed above are often simultaneously doing DG "proofs of concept." Or seeing if DM will "work for them." This is like an organization looking at double entry bookkeeping and saying "well, let's try that out in one division first."

Organizations need to be as rigorous with data as they are with inventory, suppliers, employees, or finances.

For some unexplained reason, when it is data, there is an assumption it will all work itself out. When it comes to data, which in many economic sectors is the ONLY FORM OF ORGANIC GROWTH, the data oversight is delegated to the lowest levels. The usual resourcing of a typical initial DG effort in my work over the last 10 years is under five full time equivalents (FTEs). Often it is one or *less than one FTE*. DG means learning new behaviors to replace old ones. It means a slight learning curve. That learning curve is essentially the only guaranteed cost increase from any DG effort.

Managing data assets could have been made easy 20 years ago. Now we are in a multi-trillion-dollar mess. Please do not point over to the data science folks—they are learning that most of the data they have to work with is too risky and they are starting to say the same things. As of the writing of this edition, DG is being driven more by data scientists than regulators. The data scientists are just now realizing the extent of data neglect. Many are even surprised and astonished that data was not managed better. In many places it was assumed the data could not be bad. But it is.

Also, as this is being written, organizations are investing heavily in artificial intelligence and machine learning (AI/ML). And some very hard and dangerous lessons are happening as bad data drives bad models that drive bad actions, all as the result of a deceptive or biased AI model. The growth of so-called "surveillance capitalism" is based on AI, and the backlash toward Google and Facebook is visible on a daily basis.

You cannot flourish as a data driven business or organization of any sort without some form of DG. This is a bold statement. While many organizations believe they need governance over data, few take the time to become data literate and apply this knowledge properly. As a result, many organizations start a DG program, and stumble and stop. Most failures and difficulties can be traced back to leadership not taking the time to understand more and then engage with managing data assets. Data-intense efforts are isolated, and applications areas are allowed to push back on standards. The root cause is that executive and management literacy around data and DG remains low, and programs have a hard time being sustainable. It is treated as a new, unknown program. It is treated an inconvenience. DG is not new, it is not an unknown area. *It is not an inconvenience if done correctly.*

On the positive side, some organizations are succeeding with DG and DM. The ones that succeed show measurable business improvement. "Organizations that monetize their information assets outstrip their rivals by using it to reinvent, digitalize, or eliminate existing business processes and product."[3] The returns come from data monetization, or cleaning up horrendous data quality issues, or getting their data compliant with regulators and eliminating many compliance risks.

Most of the slow adoption of DG has a very simple root cause—limited or no understanding of what it does, means, and contributes. Ironically, when I have talked to CEOs about DG, they are astounded that their firms don't have good data controls. There is a profound lack of literacy within organizational leadership when it comes to data assets. But if data is to be monetized and used to drive organizational strategy, it requires its leadership to become data aware.

Mr. or Ms. Executive—this is a call for action on your part. "Buying in" to a data project somewhere is not a solution. You need to engage in a movement toward new business capabilities.

---

[3]Laney, Douglas, "Infonomics," Gartner-Bibliomotion, 2018.

The first edition of this book assumed that some tacit awareness of DG was in place, and leadership might not know the details, but were behind it. We need to adjust this assumption. There needs to be an awareness at the highest levels that DG and its companion discipline, DM, are new, 21st century business capability areas. They are required, much like human capital, or accounting, or compliance. Imagine a CEO of a large retailer without a basic understanding of management principles, inventory management, and merchandising. That would not happen. Add DM to that list.

And for sure, most of the business and organization world has embraced the concept that data is pretty important. But the awareness needs to be reinforced with a deeper understanding of data in the higher echelons of organizations.

Most successes in DG feature strong alignment to strategy, leadership, and sponsorship. The Chief Data Officer, or similar Top-Data-Job, is a relevant and frequently necessary title.

Something to think about

Imagine you are a CEO or hold a similar position of a large organization. This means that the buck stops with you. Now imagine you get an auditor's report forwarded to you from the Board of Directors. (They do corporate governance!) The executive summary is in the form of a short letter, and it states the following:

Dear Sir/Madam:

You are well aware that as your auditor, it is incumbent on our firm to bring matters to your attention that may threaten financial progress or contain sufficient risk to cause harm to your organization.

In that context, we need to discuss your enterprise's treatment of an asset we believe is at risk:

- We have found statements in the annual report and other public disclosure that this asset is important to the organization, and represents a great opportunity.
- There are no strategic plans, initiatives, or programs linked to this asset or the aforementioned opportunities.
- Nobody can tell us where the asset sits, how much we have, or where it came from.
- Most of the uses of this asset occur at a departmental level, and there are few controls to oversee the repurposing of this asset (up to several thousand times a day).
- We have uncovered numerous instances where this asset has cost the organization material amounts, and in some cases, directly affected balance sheet and income statements.
- Many managers claim to own this asset, but an equal number try to absolve themselves of any accountability at all.
- Those that do claim ownership deny all accountability.
- For every request from compliance to destroy this asset once it offers more risk than usefulness, there are four requests to keep this asset available and "take a chance" with the risks.

Please schedule a meeting with us at your earliest convenience to review your action plan to mitigate these risk areas.[4]

The asset being addressed is, of course, data. And the functional auditor's conclusions are very typical, even though various surveys have indicated that executives feel basic data controls and oversight are in place in their respective organizations.[5]

---

[4]Original version appeared in "Making EIM Work for Business," John Ladley, Morgan-Kaufman, 2010.
[5]First San Francisco Partners, Executive Surveys, 2014–2016.

Concepts

DG is like any other governance. Someone has to mind the store while others are running the store. With data, the data scientists and analysts use the data. Systems create and change the data. But DG makes sure everyone abides by the rules. I often explain DG like this to executives:

1. You have corporate governance.
2. Due to corporate governance, your organization needs to follow reporting standards, such as HEDIS, CMS, Sarbanes-Oxley, or GDPR.
3. Due to corporate governance, your organization also commits itself to accurate reporting of financial data to shareholders, following generally accepted accounting principles (GAAP), and being audited on an annual basis.
4. The board and audit committee establish policies to remain in compliance and ensure accurate bookkeeping.
5. Neither the audit committee nor the board members actually go out and balance the books or prepare the filings. Those are operational capabilities. But there is a governance function to make sure policy is followed.
6. DG does the same thing, only for data. And you do not have that capability in your organization.

DG provides the guardrails for using and taking care of the data assets. It can include all of your documents and other content as well.

There isn't a single executive who has not received two reports, with the same data named at the bottom, and two different numbers. The initial reaction is "go back and fix this." Without DG, they CANNOT GO FIX IT. You can correct the one-off report, but the problem will remain. Whatever multiple areas compiled the data did so in their own context and priorities. Unless there is some sort of coordination and standardization, it is not going to get fixed. You will get some good arguments as to who is correct, but that's about it.

All the data scientists, Big Data, and Cloud stuff that has recently been financed will not fix the data either. Doing "stuff" with data is actually easy, if the data is useful and of the right quality. But most of the time, your data is nowhere near that useful. Getting data useful and using it to benefit the organization is the management side of data. But that is not the end of the discussion when no one is minding the store. DG is the oversight to make sure the data users and managers are using quality material and managing data the correct way.

You may have heard that your technology areas are becoming "agile," and new technology no longer requires any central control of content. Or placing it in the "Cloud" will handle DM. Those statements are only true in a narrow context, and you are unwittingly putting your organization at risk without a better understanding of the role of data in those conversations.

Think of your entire supply chain operating without any standards of quality, consistency, or timeliness. Imagine if every department at every stage in your product supply chain creates its own standards for performance. Every department counts its inventory in its own, unique way. Of course, that would be a problem. Now imagine your data in a similar way—because the reality is that your data supply chain is managed by every department doing things its own way (Fig. 1.1).

It is not hard to manage and govern data. It does not have to be upsetting. You can start small and finish big. You can unfold a DG program incrementally. You can choose, or respond to circumstances, and be as minimally invasive or aggressive as you want. You can coddle a difficult culture, or slam in a

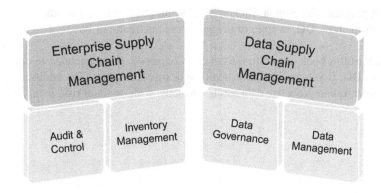

FIG. 1.1

Supply chain

compliance-mandated solution with brute force. But DG requires some new thinking, and a reasoned approach. You cannot buy it.

In some ways it is similar to efforts like Lean, Six Sigma, BPR, etc., and if your organization has done these programs you should not have an issue with DG. For some, DG may be a sea change. For others, it may not. Like these other types of programs, however, DG has clear, easily understood success factors.

Everyone goes to Lean training when it is implemented. Everyone. Implement DG, a handful go, and half of leadership doesn't show up. But like Lean or Six Sigma, executives must demonstrate they are engaged with DG if the program is to succeed. They need to learn a little about it. Delegate execution for sure. But do not delegate support and *understanding*. If data is that important, then taking care of it means you need to know a little about it. The engagement must be visible.

Let me restate this as a key point.

This author is saying DG is as significant as Lean or Six Sigma. These are new capabilities required for new times.

DG is the true indicator of the fundamental shift in how organizations view and manage data. At the end of the day, DG needs to set the parameters for consistency. It is the quality assurance to offset quality control. Deming would say it is "doing the right things," (quality assurance) vs quality control, or "doing things right."

DG is cheap. As we will cover in this book, DG does not mean an increase in overhead. It is not a new kind of IT or technology project. DG is not an accumulative program—that is, if done correctly, you do not need to add an eternally funded requirement for manpower and capital. In fact, the perfect deployment of DG will result in zero increase in costs with absolutely no visible separate DG area.

At the end of the day, DG modifies people's behaviors and business processes to think more clearly about the care and feeding of data. If done correctly, there is no need for large incremental groups of people implementing something brand new. Organizations love to jump on bandwagons and then bang on the "next big thing" until it surrenders. Organizations also have short attention spans, and shy away from new things if they seem hard. Both of these traits actually set many companies back and put them behind their competitors. When it comes to DG, the devil is in the mindset of leadership (as well as in the details).

If it is an HR program, Six Sigma, ISO, or similar initiative, it is "all hands on deck." Comply or work elsewhere. Somehow, DG, in spite of being in the same realm of business significance, gets less attention.

The remainder of this book goes into the details. Chapters 2–6 are intended to orient management and newcomers to DG. If you are intrigued, read the next few chapters. The additional awareness can only help your organization monetize data assets correctly, with minimal risk, in an ethical fashion.

Introduction

Chapter Outline

Our opinions do not really blossom into fruition until we have expressed them to someone else.
Mark Twain

Introduction

Since the first edition of this book, most organizations have embraced the concept that data is pretty important. There is awareness that data needs to be managed and governed. But programs continue to have a hard time being sustainable.

But why is it still so hard? While the main purpose of the first edition was to give the reader a solid head start to understand, obtain support, and sustain engagement for data governance (DG), we need to address the deeper factors for "doing data better." The book will still cover the deployment, implementation, or "standing up" of a data (or information) governance program. But we will add some new ideas, lessons learned, and go deeper into how to make DG "stick."

The book's core material is enhanced to cover the barriers to success more deeply and make an additional contribution to the data literacy of leadership. Lastly, the book will address reenergizing DG programs that have started, slowed down, failed to meet expectations, or even stopped.

It is also intended to supplement all other literature written about DG. For example, as of this writing, there are several confusing treatments of DG:

1. Broad announcements of Data Governance 2.0 are everywhere. I have counted at least three versions of what that means as this edition is written. Grabbing the 2.0 label might be good marketing, but it is not helping anyone. We will make sure we cover what is essential (and has always been essential) to DG.
2. There is tension between DG and service and microservice development, as well as cloud implementation of applications. These areas are not immune from the need for governance.
3. There is a growing friction between Agile development and DG oversight, while at the same time a trend to have "Agile Data Governance" is gaining traction. That will be clarified. (There is Agile DG as you will see.)
4. There has also been a tendency for less informed parties to say there are certain approaches that must be taken; that is, you need to "do DG the way expert X says it needs to be done." That is not true and will be addressed as we cover how to determine the approach required for your specific situation.

There is still plenty of advice in the following pages even if you have a program in place. In upcoming chapters, every attempt was made to keep the positions and processes disclosed as neutral as possible. There are many deployment philosophies around DG. All are "correct" if the context of the approach has been considered and will offer success in some form. In the many years of my DG practice, I have found that the common aspects of DG that cross all industries, technologies, business models, etc. are far more important than any of the differences between DG for a regulated company or a deregulated company, or DG for artificial intelligence (AI) vs master data management (MDM). However, extracting the maximum value from data requires a deeper consideration of what exactly DG means and needs to accomplish within any specific organization.

The content in this book represents what I have been doing in DG and data management activities over the years. At times I will use a different pronoun and say "we." A lot of experience and refinement has gone into the material you are about to read. These processes are not the ramblings of one person as to what should be done. This material is battle-tested. Some of the material may vary from other published methods. Where this is the case, I try and point it out and give credit where credit is due. If you need to understand the rest of the book's components, read the rest of this chapter. If you are ready to dive in, flip to the next chapter.

There is a secondary purpose to this book, and that is to provide the tools for planning, oversight, and usage of an organization's data for the maximum benefit of that organization. It is easy to forget that DG is part of a larger picture, so I will point this out when required. This book is also intended for multiple audiences, so you may see a topic repeated in later chapters in more detail than an earlier chapter.

This book will present the important, even vital, background, definitions, and preferred practices that make DG successful, no matter where deployed. It will also present a generic version of the steps and activities required to deploy DG. The case study examples and artifacts will help tie the process together. We will present examples and feedback from other practitioners.

There are templates included in the appendices as well that serve as starting points for the various deliverables and artifacts that you may need to create, or as supplements for existing programs that may not have addressed all the necessary factors required for success.

Where necessary, we will point out where different applications of data management and use call for different emphasis on the various capabilities that attend a DG program. But we will not, nor do you need to say, there is "DG for Big Data, DG for Advanced Analytics, etc." DG, and you will hear this often, is by its nature an enterprise-level thinking process. Therefore, saying there is DG for one technology and DG for another actually creates an obstacle toward successfully extracting all the value you can from your data assets.

At the end of the day, DG modifies behaviors and business processes to think more clearly about the care and feeding of data in all of its permutations and uses.

In addition, I will address some new areas that have cropped up. This includes Big Data, AI, machine learning, analytical models, cognitive systems, and supporting technologies for DG, like new tools and Graph databases. There is the rise of the data-driven organization, data monetization, and the appearance of the new roles of Chief Data Officer and Data Scientist. I also will address the actual operation of a DG program in more detail. As more DG programs mature, business-as-usual operations have presented some learning opportunities. This leads into another new topic—that of measuring DG and the value of data in general. A concept called Data Debt will get significant treatment throughout, as will as the new field of Infonomics[1] and how that interacts with DG.

---

[1] Laney, Douglas, "Infonomics," Gartner-Bibliomotion, 2018.

I will also introduce a shift that has evolved in how I approach DG efforts. To do so we will talk a lot about capabilities vs processes or policies. DG is a new business capability, so we will treat it the same way the implementation of any other new business capability is treated.

The aforementioned audiences for this book, leadership and execution, set the parameters for how it is arranged. The book is assembled in two layers. The next four chapters (3 through 6) can be considered an overview, suitable for organizational leadership. The intent of theses chapters is to raise DG literacy among leadership. The chapters present an effective executive-level overview of deploying DG so a CxO has enough confidence to hand the book to a subordinate with instructions to develop a plan of attack. The tone of these chapters is at higher levels of business conversation. We cover DG deployment from a conceptual, logical view and physical view. Often, during speaking engagements, I will tell readers to tear out the first six chapters and hand them to an executive. They will stand on their own.

The next audience addressed is the practitioner. The remainder of the book provides the details to move forward. In this way, a project manager can read the book from start to finish, but a senior leader will also find value by reading Chapters 1–6.

Lastly, each chapter concludes with some essential, focusing questions and poses additional scenarios so that the chapter material can be further discussed and examined. This is done to support the third audience of the budding practitioner or student as more and more universities are offering DG and management classes.

Therefore, while this book may seem to be a simple "how to," it is also unabashedly a treatise to convince organizations to think differently about how to manage their information and data universe. To be clear, real DG requires that organizations act differently with regard to their use and management of content, meaning data, information, documents, media, etc. Sometimes it seems too abstract or overwhelming. You have to establish a program that will oversee vast segments of your organization. Data is more and more the fuel of business, not a lubricant. DG oversees the management of all instances of data content, as well as projects and processes that create, use, and dispose of content. But you *do that already*. So this book is really designed to make this daunting change much more straightforward.

DG is absolutely a mandatory requirement for success if an organization wants to achieve MDM,[2] build business intelligence, do analytics, be "data driven," improve data quality, and/or manage documents. However, DG is not an eternally lasting add-on process. This may seem contrary to much of the literature flying about the information industry. But industry promotional literature comes from stakeholders with a vested interest in uncertainty and fascination with new things. There are many articles, for example, on how to design the DG "department," when you are really designing a framework to govern. Or how a "tool" is required to achieve success. In fact, I will describe where DG can create enormous value with only one "official" DG person in place, and absolutely no investment in tools.

As stated earlier, the next few chapters form an executive-oriented section. The purpose is to provide background, value proposition, and business relevance.

Chapter 3 will address the data literacy issue. This is an important concept. The chapter will start with the essential concepts and philosophy required for DG as well as establish a common vocabulary. Several essential concepts have evolved that require a business-level understanding.

---

[2]If you are unfamiliar with terms like MDM, data quality, and so on, stand by. We will define these in the next chapter.

As for terminology, my practice in this area has determined that the slightest variations in semantics can become huge obstacles. Therefore, we will present an essential set of terms and definitions as well as context. We will always provide the context of the term as well as refer to the definition. That way, if you read another version of a term like "policy," you at least have a frame of reference.

We will also stick to business terminology. If there is a technical aspect of a topic, it will be presented in business terms. If there is a business metaphor to lock in a point, it will be used in place of a technology metaphor.

Once we establish the terminology, we will cover the basic elements of the DG program. We will present the core managerial and business concepts required for building and operating a DG program. Since DG is a business program, you may feel quite at home reviewing the various pieces and intersections of people, processes, and information technology. Your first take away from Chapter 3 needs to be that DG is *not* part of information technology's job description.

Please thoughtfully read the text that addresses the *scope* of DG. One of the most critical errors that can be made while designing a DG program occurs when an organization has the initial conversation on scope and priorities. This examination also segues into a discussion on the business role of DG. The value proposition of DG needs to be clearly understood by executives if DG is to be successful. Finally, this part of the book is important because if DG is misunderstood, it leads to a tendency to jam it into another box on the organization chart of the IT department, and this is often fatal to the DG program.

Chapter 4 addresses the most common question when DG is getting started or is gathering new participants. "What does it look like?" The *elements*, *scope*, and *business* role sections are part of an overall segment that provides an overview of the entire DG program. It continues with a detailed examination of who should do the governing, what activities they need to perform, what is actually governed, and what DG looks like when it occurs.

Chapter 5 talks about the value proposition of DG. Very often clients will ask for assistance in developing a return on investment (ROI) for a DG program. In most organizations, the largest obstacle to starting DG is the selling—or a business case. This chapter will cover tangible and intangible business drivers for DG. Frankly, developing an ROI for a program like DG is usually done to accommodate a lack of understanding of what DG means, a lack of literacy about the value of data, political posturing, or plain old resistance to anything perceived as "new." DG is not a "project" that will grant a traditional return. DG does add value, and stating this as part of a business case is about the best way there is to frame its value proposition. We will also leverage the chapter on the business case to learn how to identify the metrics we will use to sustain the DG program.

Key concept

As you read, you will occasionally come across a highlighted section (like this). These will be labeled "Key concept," "Helpful hint," or "Success factor." They are there to reinforce the author's point, either through highlighting a point, presenting an actual interview I did for this book, or presenting an anecdote. For example, the reason that the business case for DG is not traditional lies in its nature. Justifying DG with an ROI-type calculation is like asking your accounting department or even your governing board of directors to justify its existence every year with a stated rate of return tied to a cash flow. You are attempting to justify something in a way that is inconsistent with how it operates. Then again, there is an appeal to the idea of a board of directors justifying itself with an ROI from time to time!

Chapter 6 presents an overview of the capability-driven process to deploy the DG program.

It is important to understand Chapters 5 and 6 plus the context of the concepts from Chapter 3. If you want to dive into the list of tasks to get you from point A to point B (Chapters 7–12) go right ahead, but you will end up returning to Chapters 3, 5, and 6 to figure out why you are being asked to do certain things at certain times.

Chapters 7–12 review the details of similar areas of activity we use to deploy DG: activities, tasks, work products, and artifacts. To the extent space permits, we present examples and ideas for how to execute the activities. Please understand at this point that a book like this can easily swell to 500 or more pages, so we need to strike a balance between education and writing a cookbook. Also, while the material appears to be linear, it is not. A keen observer will notice that each step in the process really takes the prior step into a lower level of detail. Agility in the DG deployment process is critical, and Agile thinking has been inserted into the process in many places.

Key concept
The first edition mentioned that the process and steps presented were not necessarily linear and could be combined into whatever approach was suitable for your organization. Apparently, I did not spend enough time with that concept. Soon after the first edition I began to hear "DG John's way" as expressed in the book vs "DG someone else's way." So, the entire process has been turned into more of a flexible inventory of activity, or a checklist, and will present examples within the various work areas of using the activities for low profile vs Agile vs central control environments.

Please note that the material will focus heavily on managing the behavioral and organizational changes required for DG. In this edition it has been expanded. This is not a change management textbook but will delve heavily into those types of activities in the context of DG. Do not take them lightly. If you do not manage the changes associated with DG, you will fail. This point has been proven numerous times since the first edition of the book.

Chapter 9 now includes a newer review of the technology for DG, where we will cover what kind of support various technologies (such as workflow, enterprise architecture, modeling, collaboration, content management, and others) can provide.

I summarize everything in Chapter 12. Under the mantra of "tell them what you are going to tell them, tell them, then tell them what you told them," I will cover a handful of mandatory takeaway concepts. In addition to the usual list of critical success factor-type bromides, you will find a lot of bullet points you can use for marketing and sustaining your DG program.

All chapters will be reinforced by a case study that weaves throughout this book.

Fig. 2.1 presents how the scope of the book tracks to a standard enterprise architecture framework, a modified view of a common framework used by enterprise architects and planners to keep track of where work needs to be done. It is called the Zachman framework (after the guy who thought it up), and many thanks to John Zachman for allowing us to use it. It is an effective means to explain how an enterprise needs to link conceptual thinking to physical implementation, which is why we included it.

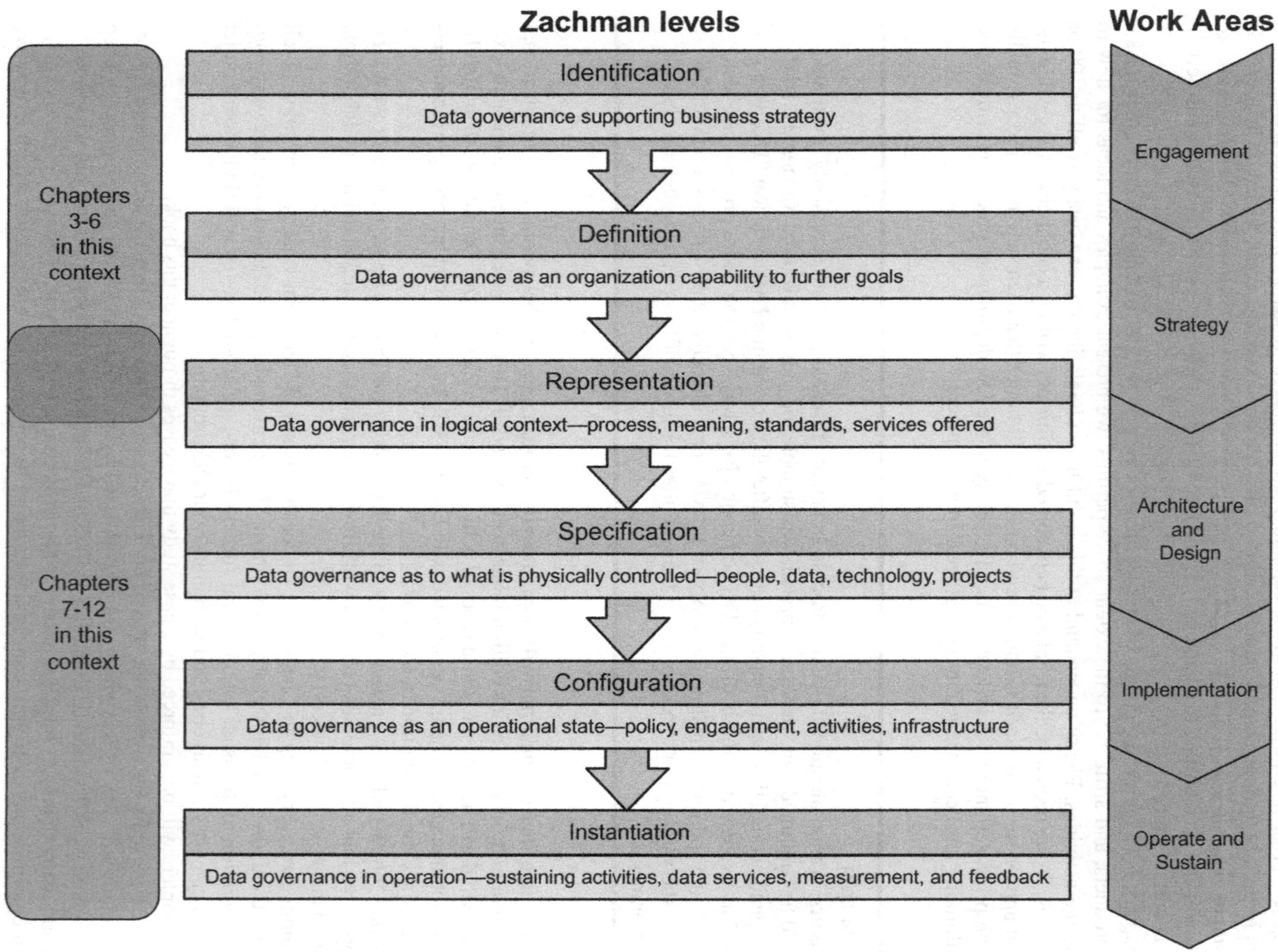

Work Areas

- Engagement
- Strategy
- Architecture and Design
- Implementation
- Operate and Sustain

Zachman levels

Identification
Data governance supporting business strategy

Definition
Data governance as an organization capability to further goals

Representation
Data governance in logical context—process, meaning, standards, services offered

Specification
Data governance as to what is physically controlled—people, data, technology, projects

Configuration
Data governance as an operational state—policy, engagement, activities, infrastructure

Instantiation
Data governance in operation—sustaining activities, data services, measurement, and feedback

Chapters 3-6 in this context

Chapters 7-12 in this context

FIG. 2.1

The scope of this book via the modified Zachman framework

It is my fervent hope you find value in starting and sustaining your DG effort within these pages. If you already have one, I hope you find some good tidbits in here to give you some ideas and make your success sustainable. If you have any ideas or feedback, please drop the author a note. Thank you for taking the time and energy to read this book.

Essential questions

Starting with this chapter there will be a list of questions for discussion and reflection. They will help associate the main points of the chapter with your own organization's characteristics.

1. Can you start DG as a delegated project without some awareness by leadership?
2. What are the differences between Data Governance 1.0 and 2.0?
3. Friction between different views of DG is a good thing. True or false?

Data literacy and concepts

3

Chapter Outline

> *Metaphors are hard to implement.*
> **John Ladley**

The importance of concepts

An organization's leadership can successfully approach data governance (DG) in two ways. They can embrace it as part of the process to get to monetization of data assets, plunge into artificial intelligence, or lower costs, and therefore support the capabilities required for that to happen. The second way is to set a vision for an organization with better managed data, authorize the necessary capabilities for the various steps that will be taken, and then let subordinates work out the details. Either way, there is a mandatory set of concepts in which organizational leadership must be made literate. To be frank, 10 years ago it would have been adequate to just let the subordinates work it all out. But data is now such a pervasive and mandatory aspect of organic growth that leadership needs to be more than just aware, they truly need to develop a solid level of understanding of those mandatory data concepts.

    While this chapter is about DG literacy, it is much more than a glossary or rehash of now common DG clichés. We need to spend some time on the deeper concepts behind the definitions of common terms. These concepts deeply influence the progress and sustainability of DG programs. Also, rather

Data Governance. https://doi.org/10.1016/B978-0-12-815831-9.00003-5

than present a definition and just let it sit there, we will talk about how the term or concept fits into practical DG practice. In addition, wherever a term or concept is being used in different ways in the real world, we will point out the differences.

We will determine a uniform definition of terms you will need to know to get through the remainder of the book.

Before you skip this chapter, saying "I get it," please reflect that a very common barrier to success of DG starts with the impression that it is a "new department," or a new means to fix data issues. Far too often I have heard an executive say, "Data stuff—oh, that goes to data governance to be fixed." To let this impression go uncorrected is to ask for a mountain of problems at a later date. So, this is an important chapter.

Data is an asset

As stated earlier, 21st century organizations need to manage data as an asset. But what does that mean? "Information is an asset" is an extremely common statement, and probably the most common information principle published within organizations. The subsequent explanation is that assets are *managed,* so *information* has to be managed.[1]

For DG to work, "asset" has to be more than a metaphor. While many experts discuss data value appearing on a balance sheet (Laney), there is a long road before accounting methods catch up to that level. You also need to look at the liability side of a balance sheet when discussing data and information, because it can hurt as well as help.

The "value" of data appears when it is used, such as in making a decision. Conversely, the negative value of data happens when data is used incorrectly or is incorrect when used. All other data activities are essentially sunk cost.

DG plays a key role in the definition and treatment of data assets. So for this book, "data as an asset" means data CAN be used as an asset through DG, ensuring its proper treatment.

Asset treatment means DG goes beyond watching projects do cool things with data or cleaning up a tactical issue. Small victories are good, but eventually the organization is overwhelmed. A colleague of mine calls this "data whack-a-mole," referring to an old carnival game. This type of cultural data fatigue costs organizations trillions per year.[2]

Data governance and governance

If the concept of managing information assets in a formal manner is accepted, we need a process to ensure that management actually takes place—and is being done correctly. Unplug your technology thinking and turn on your accountant thinking. Accountants manage financial assets. Accountants are governed by a set of principles and policies and are checked by auditors. Auditing ensures the correct management practice of financial assets. Principles, policies, and auditing accomplish for financial assets what DG accomplishes for data, information, and content assets.

DG is defined in the Data Management Body of Knowledge (DMBOK) as "The exercise of authority, control, and shared decision making (planning, monitoring and enforcement) over the management

---

[1]Ladley, John. "Making EIM Work For Business," 2010, Morgan Kaufman.
[2]Bad Data Costs the U.S. $3 Trillion Per Year, Dr Tom Redman, Harvard Business Review, 2016 https://hbr.org/2016/09/bad-data-costs-the-u-s-3-trillion-per-year.

of data assets."[3] In turn, governance is defined as "The exercise of authority and control over a process, organization or geopolitical area. The process of setting, controlling, and administering and monitoring conformance with policy."[4] This definition is, of course, roughly synonymous with government.

Slightly different definitions are often stated with an emphasis on the policy and programmatic aspects of DG. An example of one used in my consulting work is, "Data governance is the organization and implementation of policies, procedures, structure, roles, and responsibilities which outline and enforce rules of engagement, decision rights, and accountabilities for the effective management of information assets." Regardless of style of definition, the bottom line is that DG is the use of authority combined with policy to ensure the proper management of information assets.

Starting about 2015, I began to use a shorter definition to avoid controversial words like "accountability" when the situation became tense. "Data governance is a required business capability if you want to get value from your data."

That definition derived itself from another good metaphor—the supply chain. I mentioned this in Chapter 1. When a product is assembled, shipped, distributed, and then consumed, it moves through a supply chain. Supply chain management, or logistics, are well-thought out fields, with plenty of engineering and standardization. Data also moves through a parallel supply chain within an organization. In fact, when we build data architectures, we very often use logistics-derived methods to do so. DG ensures efficient design, standardization, and operation of the data supply chain (Fig. 3.1).

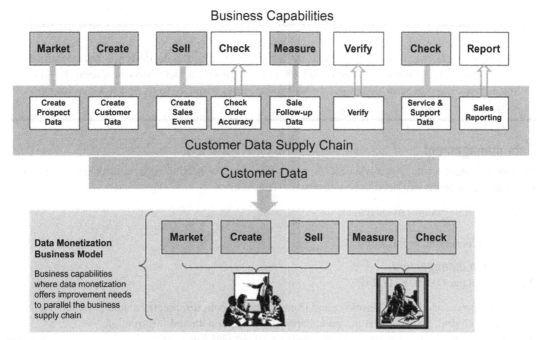

FIG. 3.1

Data supply chain

---

[3]DMBOK, 2.0, DAMA Publication, 2017.
[4]Ibid.

Regardless if the discipline is financial, supply chain, or data, there needs to be a separation of duties to ensure proper adherence to standards and policy (Fig. 3.2).

FIG. 3.2

Data governance separates duties

Data management (DM) IS the data supply chain—and that is the next important concept.

DG and DM are two sides of the same coin. They should never exist without each other.

If DG makes sure DM is happening, then together there must be a label for what they both accomplish together. DG is part of a larger discipline that has traditionally been called *enterprise information management* (EIM). In fact, most confusion about the meaning of DG stems from there being slightly differing views as to how it fits into information management (IM). So, we need to go into the concepts of data management.

> **Key concept**
> Where possible, we will use the DMBOK definitions unless the definition is not contained in the DMBOK, or industry trends have obviously altered the definition of a term. Even if the author disagrees with DMBOK, we will forge ahead with DMBOK and work around any heartburn!

Data management

Now that we have a rough idea what DG is, or is defined as, we need to address three interrelated and key concepts or terms that need to be understood. They are:

- Data (or information) management
- EIM
- Data (information) architecture

Information management

According to the DMBOK, DM is:

1. The business function that develops and executes plans, policies, practices, and projects that acquire, control, protect, deliver, and enhance the value of data and information.
2. A program for implementation and performance of the DM function.
3. The field of disciplines required to perform the DM function.
4. The profession of individuals who perform DM disciplines.
5. In some cases, a synonym for a *DM services* organization that performs DM activities.[5]

---

[5]Nagle, Redman, Sammon. "Only 3% of Companies' Data Meets Basic Quality Standards," Harvard Business Review, September 11, 2017.

Within the context of DG, the reader needs to latch onto these key terms embedded in this definition:

- *Business capability*—Twenty-first-century business and beyond requires organizations to stop looking at data, information, etc. as a convenience. The proper use and handling of data is a business obligation. Since Capabilities are an oft-used management term, think of DG as a business capability in lieu of function. A capability is a WHAT; that is, what needs to be done for an organization to fulfill its mission? What needs to be done to manage data? An example of a DG capability supporting a DM capability supporting a business capability is in Fig. 3.3.

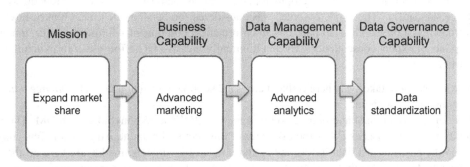

FIG. 3.3

Where data governance fits as a capability

There are distinct advantages to a capabilities-based approach:
Seven reasons why capability-based planning is effective:[6]
1. It's a top-down, whole-of-organization approach. It breaks through departmental silos by shifting from a functional view to a capability view.
2. It focuses directly on **what** an organization needs to do to execute its strategy.
3. It provides a map of the organization's overall capabilities to ensure nothing is missed.
4. It directly links initiatives and projects back to capability changes and, in turn, back to the organization's objectives. No more random initiatives that seemed like a good idea at the time, but in hindsight don't actually align to your strategy.
5. It cuts the wheat from the chaff. It helps you determine the highest priority capabilities that you need to develop, and related initiatives that you should focus on. It clarifies and optimizes business investment.
6. It stops you from jumping to conclusions about solutions too early. By delaying solution definition and doing it in the context of capabilities, it opens you up to alternatives rather than simply incrementing existing deployed equipment, processes, and people.
7. It provides a systematic way of identifying change initiatives. Many business planning approaches define mission, goals, and objectives, and then start spawning initiatives and projects. By looking at what capabilities are required to meet your objectives, it provides clarity for your initiatives.

[6]Chuen Seet, 2018, What Is Capability-Based Planning? https://www.jibility.com/what-is-capability-based-planning/.

- Program—DG is not a project with a discreet start or end point. Once initiated, it needs to operate under a "going concern" concept. In fact, over the years I have modified this stance with an additional description. Besides a "going concern" program, effective DG happens only when the organization mind-set changes around data. This is a big shift, again much like LEAN, Six Sigma, etc. These are meant to be permanent adjustments to how business is done. And other forms of governance, such as regulatory compliance, are permanent structures. DG is the same.
- Discipline—Governance, by its very definition, implies a predetermined rigor. In the early days of computer applications development, new systems analysts often asked, "How do we enforce standards?" The word "enforce" was considered too harsh at the time. Frankly, however, governance is a process that, in part, has an enforcement component—follow the rules, maintain discipline, or expect consequences. Even a minimally invasive DG effort mean formalizing the informal.

The key concept to take away here is that there is a disciplined, formal process to *manage data*. This is the beginning requirement.

IM is commonly defined and understood (via the DMBOK) as synonymous with DM. There is a community that a approaches IM as the oversight of unstructured content. All of this is fine since we have taken the position that data, information, and content (documents, media, etc.) are all the same fodder for DG. For the remainder of this book, IM, DM, and content management point to the actual management of data and content assets, and DG, information governance, and content governance all point to the same concepts and activities for oversight of these activities.

Enterprise information management
The DMBOK definition of DM or IM is generic and does require some clarification when talking about an enterprise-level program. This is because, historically, formal data or information management turns out to be a localized function. Any IT group can be more disciplined with information within a specific application or business function. However, in this book the term *enterprise information management* is reserved solely for an *enterprise*-level program. Therefore, we need to have a separate definition and concept.

Helpful hint
It is really easy for certain individuals to see the vision and relevance of enterprise DG and management. However, may others in an organization are not wired to grasp such a broad vision. One horrific mistake made by data professionals has been to thunder a message of "enterprise or bust" for everything data. While I am obligated to explain that data management and governance WILL NOT fulfill its potential without an enterprise mind set; that does not mean it is implemented in one huge, disruptive big bang and all members of the management team must convert or go away. Being a DG terrorist is as bad as choosing to remain ignorant on data literacy. Both are culturally harmful.

EIM is the program that manages enterprise information assets to support the business and improve value. EIM manages the plans, policies, principles, frameworks, technologies, organizations, people, and processes in an *enterprise* toward the goal of maximizing the investment in data and content.

You cannot deploy EIM by department. EIM represents more of the direction, philosophy, and mind-set required to manage data assets. EIM is like democracy. It is a societal philosophy. After you accept it, it is easier to work out the federation of states and such, but you all need to accept the philosophy. As defined here, information or data management represents the day-to-day "stuff" that actually has to be done to achieve the information asset management. IM (or DM) is simply the program that manages information as a recognized and formal asset. EIM is the enterprise-level support and mind-set.

Data architecture

Another term often heard within a conversation related to IM or DG is *data* or *information architecture*. The DMBOK definition of information architecture, or data architecture, is somewhat convoluted and tilted toward a technical explanation. The entire definition can be read in the DMBOK, but here is a summary:

1. A master set of data models and design approaches identifying the strategic *data requirements* and the components of DM solutions, usually at an enterprise level.
2. The "data" column of the Zachman Framework for Enterprise Architecture identifies six different classes of design artifacts, each representing a different level of abstraction. (Note: This is not exactly a business definition like we promised. See Chapter 2 for an explanation of Zachman.)
3. In some common usage, the physical technology infrastructure supporting DM, including database servers, data replication tools, and middleware.

The author would never use the preceding definition when educating management as DG is deployed. Rather, a much simpler version would be used:

- Data architecture is a representation of the DM environment, its components, and their interactions. This picture, or abstraction, interrelates the framework, people, processes, projects, policies, technologies, and procedures to manage and use valuable enterprise information assets.

The governance "V"

Make sure you do not confuse the *management* of data with *ensuring data is managed*. Let's introduce a concept based on duty separation used throughout this book, called the **Governance V.** (See Fig. 3.4.)

The left side of the V is governance—providing input to data and content life cycles as to what the rules and policies are, and activity to ensure that data management is happening as it is supposed to. The right side is the actual "hands on"—the managers and executive who are actually doing the IM.

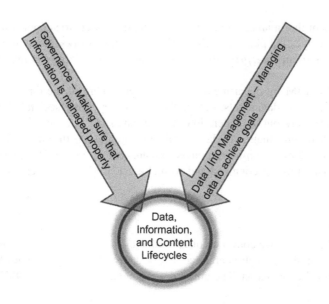

FIG. 3.4

The governance "V" model

The left side is DG, the right side is DM. It is absolutely essential that you keep this next phrase in mind all through your DG program:

DG is NOT a function performed by those who manage information.

This means you should strive for some separation of duties between those who manage and those who govern. The V is a visual reminder of this. You cannot always achieve perfect separation of duties (you run out of people). This is a key concept that business people understand, and IT staff often experience as a problem. For example, in business there are auditors and managers. Managers control, monitor, and ensure work gets done and rules and standards are adhered to. Auditors verify compliance to standards and define and implement new controls and standards as required. This is exactly the same protocol that is required by DG. The DG "area" identifies required controls, policies, and processes, and develops rules. Information managers (essentially everyone else) adhere to the rules.

At the confluence of the two lines (the bottom of the V) are the activities that operate the organization through maintaining information life cycles—creation, use, manipulation, and eventual disposal of data, information, and content.

Helpful hint

Keep a version of the Governance V around all the time—you will be amazed at how much it helps.

An internal definition of DG would take the generic definition of DG, incorporate the Governance V, and tune it to a definition more specific or relevant to an organization. For example:

"Data governance represents the program used by ACME to manage the organizational bodies, policies, principles, and quality that will ensure access to accurate and risk-free data and information. Data governance will establish standards, accountabilities, responsibilities, and ensure that data and information usage achieves maximum value to ACME while managing the cost and quality of information handling. Data governance will enforce the consistent, integrated, and disciplined use of information at ACME."

Some additional definition examples

DG is a business process separate from data (or information) management that affects the entire business. *(Data Strategy Journal, October 2007)*
DG is a framework of accountabilities and processes for making decisions and monitoring the execution of DM. *(financial organization)*

- using a horizontal perspective of the organization and focusing on the major "pain points" for our business areas. *(financial services)*
- designating people, process, and technology. *(Data Strategy Journal)*
- the orchestration of people, process, and technology to enable the leveraging of data as an enterprise asset. It affects all organizational areas by lines of business, functional areas, and geographies. *(software company)*
- using rules, monitoring, and enforcement with culturally acceptable techniques. *(Data Strategy Journal)*
- a system of decision rights and accountabilities for information-related processes, executed according to agreed-upon models that describe who can take what actions with what information, and when, under what circumstances, using what methods. *(consultant)*

To be clear, it is the exercise of executive authority over business data. *(chemical company)*

At this point, we have explained the concepts of EIM, IM, and DG. It is perfectly understandable that the reader might be thinking, "So what?" However, the time taken to review these concepts is worth it—not for a business person who is reading this book, but for an IT person. Frankly, any businessperson understands these concepts when presented in the context of "hard" assets.

So, for the sake of review:

- EIM is the program similar to supply chain management—an overall philosophy of management toward a goal of efficiency.
- DG is like auditing. Rules, standards, and policies are defined and verified. DG is the QA/audit/compliance aspect of EIM. DG designs the rules that information is managed by. IM does the managing.
- DM is like, well, DM, or IM. IM is the same as inventory management—the actual touching, using, moving, tracking, and managing activities of the assets.

Solutions

In addition to the aforementioned concepts in this chapter, there are concepts and terms you need to understand that are more related to various business solutions that DG will support. Before we look at the specific types of solutions, however, we need to understand one key theme related to all of these solutions: *regardless of the content type or technology being governed, DG is essentially done the same way.*

At the time of writing there is a lot of talk about new Data Governance, Agile Data Governance, Adaptive Data Governance, and Data Governance 2.0 (that was inevitable I guess) and probably others. Frankly, these all represent people learning what was already known.

That is, from a "how-to" DG perspective, none of the solutions we are defining make a bit of difference in how you deploy DG. You may have different priorities, deal with different groups of people, have a different operating and engagement structure, or emphasize certain DG capabilities before others, *but 80% of what you are doing is the same.*

The DM areas that require and usually trigger DG programs are:

1. Master data management (MDM)
2. Data quality
3. Business intelligence
4. Analytics, or advanced analytics
5. Artificial intelligence or machine learning

Master data management

MDM is actually a revision of another solution set that started with customer data integration (CDI). The theory was to create a "gold copy" of a crucial data subject (i.e., customer). The gold copy is the single source of truth regarding customer, and all other uses of the concept of customer must be subservient to the central or gold copy. CDI became MDM when the marketing types realized that other subjects besides customer required gold copies. Items, products, vendors, etc. are all areas where companies tend to have multiple versions, which are inconsistent or too contextual. In the old days, we called these files master files—hence, master data management.

The DMBOK states that master data is "…the data that provides the context for *transaction data.* It includes the details (definitions and identifiers) of internal and external objects involved in business transactions. [It] Includes data about customers, products, employees, vendors, and controlled domains (code values)."[3] Accordingly, MDM represents the "processes that ensure that *reference data* is kept up to date and coordinated across an enterprise. The organization, management, and distribution of corporately adjudicated data with widespread use in the organization."[4]

Obviously, if MDM represents the process to manage a category of data across an enterprise, then DG needs to come into the picture. Later on, we will talk about DG being mandatory for MDM.

DG visibly supports MDM in several ways:

1. Ensures that standards are defined, maintained, and enforced.
2. Ensures that MDM efforts are aligned to business needs and are not technology-only efforts.
3. Ensures that data quality, process change, and other new activities that are rooted in MDM are accepted and adapted by the organization.

Data quality

Data quality is probably the single most discussed term or concept in the EIM/DG universe. This is easy to comprehend once you understand what it really represents. Data quality is simply the root cause of the majority of data and information problems. Remediating data quality is one of the main drivers of DG and MDM. Many organizations are surprised when they find out how many issues there are with their data and few can pass any data quality scrutiny.[5]

The DMBOK addresses data and information quality separately. As you already know, this book does not separate the two concepts, as governance is governance for both of them. Both are presented here:

- **Data quality** is the degree to which data is accurate, complete, timely, consistent with all requirements and business rules, and relevant for a given use.[6]
- **Information quality** is the degree to which information consistently meets the requirements and expectations of knowledge workers in performing their jobs. In the context of a specific use, the degree to which information is meeting the requirements and expectations for that use.[7]

Obviously, while the two definitions are different, they are certainly pointing in the same direction. The best way to understand data quality is that the content in question has to be effective or fit for its purpose. This means if your organization feels that customer data is not of "good quality," you need to understand what purpose, action, or context is involved and how the shortfall is measured. Does bad customer data mean a wrong address or excessive duplication? You need to understand that "bad data" does not just appear, and is almost always corrected by a change in processes or habits, or both. That is why the definition of data quality appears now in this text. It is a key driver of governance, because without governance, data quality efforts become costly one-off exercises.

DG supports data quality solutions via:

1. Ensuring that data quality standards and rules are defined and integrated into development and day-to-day operations.
2. Ensuring that on-going evaluation of data quality occurs.
3. Ensuring that organization issues related to changed processes and priorities are addressed.

Business intelligence

Business intelligence (BI) has grown from a term coined by the Gartner Group in the 1990s. It has since morphed into a label that describes a self-perceived cool way of looking at data. Our DMBOK reference states BI is:

1. *Query, analysis*, and *reporting* activity by *knowledge workers* to monitor and understand the financial and operational health of the enterprise.
2. Query, analysis, and reporting processes and procedures.

---

[7]DMBOK.

3. A synonym for the BI environment.

4. The market segment for *BI software* tools.[8]

From our DG perspective, we will stick with this definition: At its roots, BI means one core concept—using information to achieve organization goals. The rest is techno-speak and not relevant to our discussion on governance. DG enhances BI in a number of ways:

1. DG is used to ensure that BI activity is aligned with business activity. Many BI-related efforts never reach potential because they merely regurgitate data back to a requestor versus trying to change the business.

2. DG ensures that data quality is defined and supportive of BI. Data profiling activity is defined in the context of supporting BI data quality, and data quality remediation is occurring.

3. DG is used to ensure consistency in data standards and algorithms. Far too often, multiple business areas define a metric with the same name and different meaning and/or algorithm.

4. Lastly, we promote DG as important to enforcing the defined BI delivery architecture (i.e., make sure that organizations avoid exponential growth of spreadsheets, Access databases, and uncontrolled redundancy).

Analytics and advanced analytics

Related to BI is analytics. Analytics is the application of modern data technologies for data discovery, interpretation, and communication of meaningful patterns in data. Analytics relies on statistics, advanced languages, and advanced math to derive new insights. Statistical models are good candidates for governance, as incorrect models can send organizations in wrong directions.

However, analytics has emerged as a synonym for BI—and other data examinations. For some, it is the process of analyzing information from a particular domain, such as website analytics. For others, it is just more complicated BI. At its essence, analytics is a more sophisticated use of data. Typically, analytics is used in conjunction with the phrase "Big Data," although both can exist without the other. analytics is important to DG, in that much DG is driven by firms attempting to monetize data assets through analytics. Often DG has to make sure the definition of analytics is applied consistently within an enterprise, let alone make sure that data quality and context does not result in incorrect results of data analysis.

Artificial intelligence or machine learning

Artificial Intelligence (AI) and machine learning extend the advanced analytics model into more of a closed loop application; that is, a system becomes self-learning. Algorithms then suggest and monitor automatic responses and actions. Again, oversight of models and their application make this corner of data management a likely area for governance. Data quality is also hugely important to AI, and governance plays a large role in ensuring legitimacy of AI models.

Other terms

A few other terms we will use frequently are related to actual elements of a DG program. We will review these in detail in upcoming sections. However, it is good to be aware of these before proceeding. We will go through these quickly as many are addressed in detail later in the book.

---

[8]DMBOK.

Principles

At the heart of effective governance are organizational principles. The DMBOK defines them as:

1. A fundamental law, doctrine, premise, or assumption
2. A rule or code of conduct

Principles are statements of philosophy. Think of them as a bill of rights—core beliefs that form the anchor for all policies and behaviors around information asset management (IAM). They are beliefs to be applied every day as guidance for procedures and decision-making efforts. Principles are not to be confused with policies (see the following) or rules.

Often, we see organizations lay out a set of rules—a blend of philosophy, policy, process, and enforcement. This is not an ideal approach; rules do not have the weight of belief, they are hard to maintain, and are inflexible. DG is a behavior change, not process revisionism. It may seem heavy handed but going about it with the structure shown in Fig. 3.5 pays off over the long term.

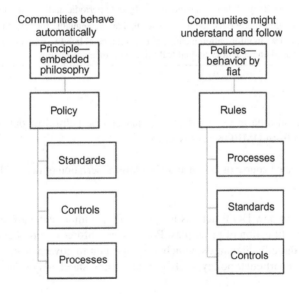

FIG. 3.5

Principles need to drive policies

Principles are significant enough that I designed an overarching set of principles that are deliberately modeled on and placed alongside Generally Accepted Accounting Principles (GAAP). GAAP and the United States Financial Accounting Standards Board set forth the essential and *mandatory* principles and standards for financial accounting. Called GAIP™ (Generally Accepted Information Principles), I urge clients to incorporate these as essential components of their principles. Table 3.1 presents a summary of GAIP™.

Table 3.1 GAIP™—generally accepted information principles

Principle	Description
Content as asset	Data and content of all types are assets with all the characteristics of any other asset. Therefore, they should be managed, secured, and accounted for as other material or financial assets.
Real value	There is value in all data and content, based on their contribution to an organization's business/operational objectives, their intrinsic marketability, and/or their contribution to the organization's goodwill (balance sheet) valuation.
Going concern	Data and content are not viewed as a temporary means to achieve results (or merely as a business by-product), but are critical to successful, ongoing business operations and management.
Risk	There is risk associated with data and content. This risk must be formally recognized, either as a liability or through incurring costs to manage and reduce the inherent risk.
Due diligence	If a risk is known, it must be reported. If a risk is possible, it must be confirmed.
Quality	The relevance, meaning, accuracy, and life cycle of data and content can affect the financial status of an organization.
Audit	The accuracy of data and content is subject to periodic audit by an independent body.
Accountability	An organization must identify parties that are accountable for data and content assets.
Liability	The risks in information means there is a financial liability inherent in all data or content that is based on regulatory and ethical misuse or mismanagement.

Policies

Policies (or policy) are an area in the field of DG that can be helpful or destructive for a new DG function. The definition from DMBOK seems simple:

> "A statement of a selected course of action and high-level description of desired behavior to achieve a set of goals."

However, it is too easy for new DG functions to spew policy without any substance. The real essence of policy is that it is a codification of principles. Policies are enforceable processes. Principles tend to be too lofty to enforce directly. Standards, which are important to governance, are a type of policy, or even a characteristic of a particular policy, such as data naming standards or data quality standards.

Some final core concepts

Lastly, there are a few more concepts that need to be understood in regard to DG. These are important to cover now because:

- DG programs often do not get started smoothly.
- They are often perceived as expensive.
- They are often scoped incorrectly.
- They contain changes to organizations that are often overlooked until it is too late.

We will go into deeper detail on each one of these but, again, it is good to be aware of them as you proceed.

E for enterprise

DG is an enterprise program. It can be implemented locally, but must never be considered as a localized project. Would you implement financial controls in one department but not in another? We need to view DG the same way.

Business program

DG is for your business or organization. It is *never* an IT program. Later, we will talk about why the CIO should never be in charge of DG. In fact, IT and technology areas are just as likely to be changed or enforced as is a business area when it comes to managing information. For now, just keep in mind that we are building a business program and that program must add value over time.

Evolution vs revolution

DG needs to be implemented iteratively, in a carefully designed deployment. You need to learn how to govern. It is not instinctive. It is not a big-bang suitable endeavor. Only the hardiest or most desperate organizations can tolerate a massive shift to governed data from nongoverned data. Look at the road to DG logically—if you are reading this book, you are not sure how to do it. You need to evolve through a process to learn how to do DG. Regardless of approach, there are four distinct stages to learning, which apply to organizations as well as individuals.

1. Rote—repeat, but not understand—the organization can express definitions of DG concepts.
2. Understanding—the organization can comprehend the nature and importance of DG (a lot of DG programs stop here).
3. Application—the organization knows enough to start to apply the concepts of DG, but only as a direct response to a trigger (e.g., data quality is poor, so we start to govern data quality).
4. Correlation—the organization can apply the concepts creatively and to more complex situations (e.g., retrofit some kind of governance to an enterprise resource planning or MDM program that has gone bad).

Information management maturity

A widespread method to view an organization's ability to execute IAM is through the lens of a maturity scale. There are as many flavors of information management maturity (IMM) scales as there are consultants and vendors providing information solutions. There is a thorough coverage of this in "Making EIM Work for Business"[9] but Table 3.2 shows a generic expression of maturity.

---

[9]Ladley, John. "Making EIM Work for Buisiness," 2010, Morgan Kaufman.

Table 3.2 Capability-based information management maturity model	
IMM stage	**Description**
Initial	The organization is entrepreneurial; individuals have authority over data, so information maturity is chaotic and idiosyncratic. Business rules or criteria for behavior are nonexistent. Data quality is far from integrated, and data handling is costly.
Repeatable	Departmental data becomes the norm. Any sophistication in usage—such as analysis—is departmental, specialized, and costly.
Defined	The organization starts to consider an enterprise view, and looks for some sort of integration across applications and silos. A desire for data accountability evolves. Strategic alignment to the business becomes an activity in IT. Standards are developed, and data quality becomes formal and may centralize. Data usage becomes more common, and efficiency of DM improves.
Managed	Data and content assets are tracked, lineage of all content is understood and documented. Analytical results are used to close process loops. Emails, documents, and web content are also managed, and can be called up alongside "rows and columns." Data quality is built into processes instead of being corrected post facto.
Optimized	There is no need to determine if information assets are managed effectively—they are woven into the fabric of the organization. There are effective measures in place to allow IM to support business innovation. The organization can place a value statement on its content, if not the balance sheet.

IMM is a key concept in that it represents a broadly understood means to measure the progress and effectiveness of DG. If IMM improves, DG is working. It is not a report card, merely a measure.

Don't confuse maturity with the levels of learning. They are not interchangeable. They support each other. Depending on your organization's culture and environment, you may possibly need to execute all four layers of learning to get through each maturity level.

Things will change

The reason you are reading this book is that something is amiss with your data. By definition, if something is wrong, it needs to be fixed. Fixing anything means making a change to ensure that the fix is never needed again. The bottom line is that DG is not done with an expectation of "business as usual" across your business and technology functions. There will be changes. Some of them will not be well received. Part of deploying DG means managing changes.

Summary

DG is a key element of managing data assets. I have contrasted DG with data and information management and reviewed the specific solutions that may trigger DM and DG. The relationship of DG and DM is key in understanding the role of DG and keeping all of the terminology straight.

Managing data as an asset describes a business-based approach to ensure that data, information, and content are all treated as assets in the true business and accounting sense—avoiding increased risk and cost due to data and content misuse, poor handling, or exposure to regulatory scrutiny. Please go back and review that sentence. Applying DG means treating data as an asset, but not in a metaphorical sense. We truly mean as a real business asset. You may not see your "information value" on a balance sheet, but to be certain, if you view data asset management in the true business sense, deploying DG is a whole lot easier. It becomes a necessary business capability. To be clear, if you are serious about governance or any of the solutions that require its application, you are committing to data asset management. Think of another corporate or organization asset that can function without:

- Standards of use
- Accurate financial tracking
- Statement of value to the organization
- Assignment of accountability and responsibility

An asset requires standards, tracking, value, and accountability. EIM, DG, MDM, and all of the other concepts listed earlier, exist to manifest the management of your data assets.

Few of the organizations will view these concepts and terms as a uniform discipline without embracing some fundamental changes in their view of data. Yet they all want "data governance" to be implemented. They want to manage information as an asset. We usually discover they do formal IAM in pockets, but never extract maximum benefit of sustainability. Often the projects related to the various pockets of solutions fail. We can always tie the failure back to not adopting the right mind-set. The organizations doing isolated data effort go through the motions, hire consultants, and buy the right tools. However, they fall short when it is actually time to change the day-to-day treatment of data, information, and content. The solutions do not fully work unless you start to think in terms of data assets. Therefore, manage data as an asset. This is the crucial mind-set—the overarching philosophy. The elements of DM and DG provide the framework (remember the V) that ties the participants together, but clearly delineates a system of checks and balances. What they accomplish together is truly managing data as an asset.

Essential questions

1. How is DG similar to accounting concepts?
2. There are so many types of data technologies that you need different governance for each one. True or false?
3. Why is discipline a part of any DG effort, regardless of approach, such as noninvasive or very visible?

Overview: A day in the life of a data governance program and its capabilities

4

Chapter Outline

> *Laws are sand, customs are rock. Laws can be evaded and punishment escaped but an openly transgressed custom brings sure punishment.*
> **Mark Twain**

What does it look like?

A data governance (DG) program really has one clear goal—to create a common place business-as-usual program that ensures your data does better things rather than harmful things. After that happens, any impression of DG as "special" or "new" should disappear. Let me rephrase that. The long-term goal of DG *is to disappear into the everyday operations of an enterprise*. It needs to be institutionalized, like Lean Six Sigma or any other change in work behavior or business capability.

This applies to any technology, activity, or location along the data supply chain that requires DG to be successful—advanced analytics, data lake, artificial intelligence (AI), data integration and movement, external data acquisition, etc. There should never be an isolated DG for analytics or DG for AI, for example. Remember almost 80% of your DG capabilities are applied uniformly across all types of data management and usage. What varies are:

1. Implementation style
2. Federation
3. Specific capabilities

Data Governance. https://doi.org/10.1016/B978-0-12-815831-9.00004-7

A frequent question I get from executives is "what does it look like?" This means that the current understanding is DG is something entirely different and unrecognizable.

Ensuring a good understanding of how a DG program looks and works is essential to getting participants engaged. The concept of assimilating DG into everyday corporate life adds additional challenge, since you are not only defining and implementing a discrete program; you are also attempting to alter behavior to a point that the long-term program is visible only through verification and adjustment.

Whether DG is new or has become endemic and institutionalized, there is a collection of elements that characterize a DG program. Understanding how these aspects work together aids in understanding the "big picture." This chapter reviews the scope and content of these elements at an executive level, it covers what DG looks like in operation, and it addresses the types of capabilities that can be deployed. The chapter will start with a short overview of a case study.

Introducing the case study

Rocky Health Systems is a regional provider of medical services in a large western state of the United States of America.[1] It serves a rural population and has grown over time by various small hospitals and physicians merging and remerging. As such its applications are fragmented.

Rocky started a DG program one year ago. It was initially driven by a joint request from the CFO and CMO (Chief Medical Officer) when a large reimbursement ($US 5 million) from a government agency was withheld due to a report where "the data was wrong." There was a lot of finger pointing as to how that happened, then a claim that the data was, after all, correct. But when asked by the agency to prove its correctness, Rocky was unable to present sufficient proof that the data was, in fact, representing the condition that was reported.

A small group was tasked to investigate the data issues on the report, and, as might be expected, uncovered a large amount of pent-up frustration about almost all of the data within the organization. This was compounded by a mandate from the new CEO that all areas start to hit specific targets laid out in a large series of metrics. Development of a corporate scorecard was initiated. No one agreed on the resulting metrics, and leadership was afraid to manage them.

Eventually, someone said "we need to fix the data," and the DG program was initiated. A consultant was called in but only on a limited timeframe and budget. Tight budgets and a hesitancy to share problems with others constrained the DG effort. The decision was made to start with a low-profile DG program: do DG, but on a limited scale, being sensitive to any impression of higher overhead or excessive bureaucracy.

Since "fixing the data" was front of mind with most stakeholders, DG had trouble getting traction as the party that pointed out *how* to fix the data, but did not actually fix it. In addition, the entire DG team consisted of one full-time manager in the compliance area and a part-time analyst assigned from the business intelligence team.

The low-profile program started to collect data issues. A data issue log was initiated. The DG side of the "V" was the compliance manager walking around, socializing the program, and reexplaining the issue tracking. Once issues were logged, the compliance manager made sure that IT put a resource on the fix. Luckily, there was a DG sponsor (the CFO) who was also the direct boss of the CIO. So, the right side of the V was kept engaged. Data quality issues were addressed, and report results improved. Six months after the program started there was no issue with the reimbursement from the government agency, and the current and the back amounts were paid.

This success was publicized by the DG team via the sponsor, and the leadership team gave the thumbs up to bring a consultant back in and help Rocky expand the program. A set of DG metrics were defined to enable progress tracking, and additional resources were placed on a new DG council. Rocky Health continues to implement DG capabilities with a low profile approach—fixing new problems with additional DG capabilities or expanding existing capabilities.

[1]Rocky Health Systems is a fictitious organization. It is based on an amalgam of several healthcare clients, and features their collective data problems, as well as problems from different industries that were worthy of covering in the case study. Any resemblance to persons alive, dead, fictional, or real, whether you like them or not, or are even related in a tiny bit, is purely a coincidence.

Notice there was not a big effort to accomplish improvement in data assets. In addition, there is a degree of continuity with DG now. This is the result of clear planning, addressing the potential organization issues, and finding a use case where the value of data management and governance can be made clear.

The scope of data governance and data management

We already mentioned that DG is an enterprise concept. There needs to be an acknowledgment that the organization will adopt a global mind-set requiring greater rigor as far as handling its data and information. DG must always be viewed as an enterprise approach. However, declaring the scope of DG is a bit more complicated than saying, "We are governing everything!" It means considering some key factors affecting scope, and then making sure you are very clear as to the definition of DG's reach and span in light of these factors. This is the first area you need to understand when learning how DG "works."

The four factors to consider that affect the scope of DG are:

- *Business model*—The type of organization, its corporate hierarchy, and its operating environment.
- *Content being governed*—The type of content (data, information, documents, etc.), its location, and its business relevance.
- *Degree of federation*—The extent or intensity by which different content is governed.
- *Development methods*—The manner in which an organization develops and maintains databases and application can drastically affect scope.

Business model

For example, a large multinational company does not have to deploy a global DG program from the initial mention of the word governance. The scope can be a self-contained line of business. Suppose you are a large international chemical company. Your business model may contain pharmaceutical, agricultural, and refining divisions. Each of these would operate on a self-contained basis. You may then have three DG "programs" that are each similar in makeup, but separately accountable.

Then again, what if you are a global retailer with a tightly woven international supply chain? Your business model is the same across all countries, but you also have some product variations and regulatory areas to consider. The scope of your DG is most likely global, but with some careful consideration of federation.

In Fig. 4.1, Company A is a large multinational organization, but all regions share its data and content. DG would, ideally, be applied across the entire entity. Remember, applied is not the same as implemented. You may implement in a low profile, less invasive manner, or choose a more aggressive approach. Either way DG is implemented gradually, but the vision is enterprise-wide. Company B is a large company as well but has several very distinct business units. They do not share common information, so in this case DG can be implemented or applied by business unit.

Company A

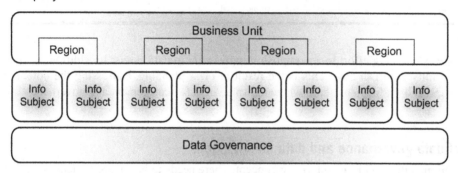

Company B

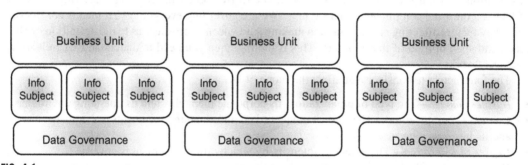

FIG. 4.1

Data governance scope

The detailed business model is also an area to consider when scoping DG. DG will often require a change in business processes. A typical example is when an organization is addressing data management of a core data area, or domains. Typically, this is Customer Data or Item Data. A new data capability called master data management is implemented.

It is very often the case that the operators entering data into many applications have also maintained the data area in question. Going from many item master files to a single, truthful source of items is a typical example of where day-to-day business processes need to change. The scope of DG needs to clearly mention this possibility.

In addition to the DG scope being dependent on the business model, it can also be dependent on type of content.

Content

This book does not consider a distinction between DG and information governance. For the most part, you do not govern different types of content differently. At the end of the day, the activity to govern business intelligence data, operational data, e-mails, contracts, documents, analytical or machine learning models, or even media is driven by the same reasons and entails mostly the same capabilities.

However, we do need to be clear within a specific organization what types of content are subject to DG. Certainly, master data, business intelligence data, and other forms of structured data are most likely governed. However, a highly regulated company may also need to govern e-mails and contracts.

A company where safety is a major issue may need to place its governance focus on guidelines and procedures. A government body may need to zero in on governing access to public documents and interagency data exchange while protecting individual privacy.

The types of content subject to DG will heavily impact where the DG program resides, who holds accountability, and how the organization deploys the DG program. It will also influence the types of tools and policies that the DG organization must define.

Content types are also important. This influences the capabilities required and the detailed governance processes. Differing content types will have unique life cycles. For example, content that is a structured type of data, like a transaction, may come and go within a fiscal year, and governance will tend to focus on the usage of that data within the time period. An unstructured type of data, like contracts and e-mails, may need to be kept for decades and may be subject to legal discovery or strict classifications of privacy or privilege. Obviously, there will need to be consideration of the details of governing these different types.

Helpful hint

If you want to see a modern example of the need for governance and a precise definition of scope, look no further than your own local SharePoint™ or Notes™ repositories. It is hard to find a better example of vital content getting neglected and descending into expensive repositories of data decay and "garbage dump" manifestations. The aforementioned tools offer excellent collaborative capabilities but become nothing more than document graveyards where old Word documents go to die. They remain expensive but harmless until a legal case pops up and the company discovers it should really have deleted those items a long time ago.

Data lakes in big data environments are a close second. Large investments are made in big data technology and only a few people know where anything is, or what it means. For many organizations a total and unequivocal lack of oversight created an enormous, expensive corporate data risk.

Third is external data acquisition. As data becomes a form of currency or fungible good, many organizations sell or exchange data. Companies gleefully engage in contracts to acquire external data with no vetting of usefulness—just accepting of the abstract description of the product and throwing it into the data lake. Again, the result is huge risk, more cost, and failed expectations.

Development methods

The development and maintenance of applications and systems should be accounted for in the type of governance as well. If the CIO oversees all applications development, then leadership must address that another body will have oversight over the CIO. Sometimes this does not play well with the CIO.

Many organizations have a defined development process, or systems development life cycle, for defining and deploying automated systems. Some aspire to an Agile delivery environment. Others have more traditional "plan and build" approaches. Sadly, many organizations have no formal statement of how to get applications and technology implemented.

Few of them have built any type of consideration for designing and delivering around DG policies and standards. Very often, DG will need to write the enhancements to corporate IT application development when structured information is being governed, or work with compliance to oversee unstructured content. The enhancements take the form of updating traditional project artifacts, and branch out to additional tasks, or new approvals and checkpoints. When unstructured information is subject to DG, we often must modify workflow and document management policy.

Federation

One of the most important concepts affecting the nature and scope of DG is that of "federation." We covered this definition earlier.[2]

For DG, this means defining an entity (the DG program) as a blend of governance capabilities that touch various functions in the organization. The federation of a DG program is a definition of where and how standards will be applied across various layers and segments of an organization. Politically, the United States is a federation, an organization of states with a federal oversight layer. In the United States, some activities of government are central. There is a central military and reserve banking system. Other functions of government operate at the state or local level, such as medical care and law enforcement. A DG program will necessitate the same type of definition of the required layers of governance functionality.

The definition of federation will influence the operations of your DG organization, its processes, and principles. Note in Fig. 4.2, I show a heat map where similar data assets can be tightly governed (in the center or hot zone), or more loosely governed (on the fringe or the cool zone). The solid areas indicate a governed area called "item," where there is tight control of global items, slightly looser control on regional items, and local items are barely governed. The dotted areas point to another subject, "customer." There is still the tight control for centrally used customer content, but the regional and local are treated the same. So, the federated intensity of DG differs by content type.

Scope factors that affect the federated layers and activities are:

- *Enterprise size*—Obviously, huge organizations will need to federate their DG programs, and carefully choose the critical areas where DG adds the most value.
- *Brands*—Organizations with strong brands may want to consider this in their DG scoping exercise. One brand may need a more centrally managed data portfolio than another.
- *Divisions*—One division may be more highly regulated, therefore requiring a different intensity of DG.

---

[2]The *Webster Dictionary* definition of *federation* offers some insight: *an encompassing political or societal entity formed by uniting smaller or more localized entities: as a: a federal government, b: a union of organizations the act of creating or becoming a federation; especially: the forming of a federal union.*

- *Countries*—Various nations have different regulations and customs, therefore affecting how you can govern certain types of information.
- *IT portfolio condition*—An organization embarking on a massive overhaul of applications (usually via implementing a large SAP or Oracle enterprise suite) will have definite and specific DG federation requirements. When a DG effort is getting started, it is usually understood (at some intuitive level) what the condition of the application portfolio is. A portfolio undergoing radical updating is a prime target for DG, and the impact on DG scope needs to be clear.
- *Culture and information maturity*—The ability of an organization to use information and data is referred to as its information management maturity, or IMM. The way an organization gets its work done is usually called culture. In combination, the specific IMM and culture of an organization will affect the scope and design of the DG program. For example, an organization that is rigid in its thinking and has a low level of maturity will require more centralized control in its DG program, as well as encounter more significant change management issues.

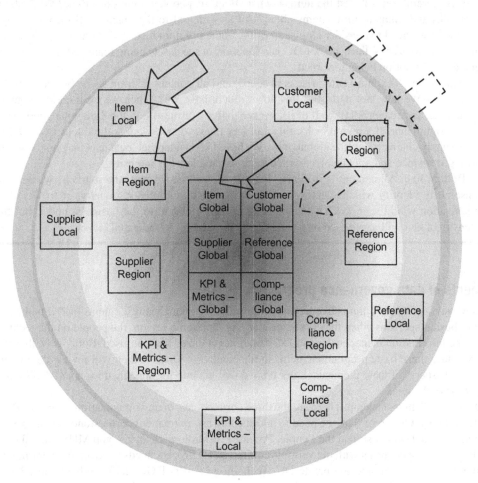

FIG. 4.2

Federation "Heat Map"

Short case study

Don't fall into the scope trap of identifying the scope of DG with size or market dominance. You need to rationally consider influencing factors we have presented; that is the business model, the assets to be managed, and what type of federation is required. Using Fig. 4.2 again, let's assume it represents a global retail organization, with many brands, but the same distribution and merchandising network.

Business model—the business model is global, with heavy dependence on economy of scale across the supply chain. So, our scope will lean toward the entire organization—we will not be excluding any functions, like merchandising or warehouse.

Content being managed—obviously there is a lot of content in a large organization, but consider the variety—retail is, at its core, pretty simple. You buy stuff from one place and sell it to someone else. The main content is anything used or descriptive of the "stuff" and getting it sold. Be careful—it isn't just the items—what about the people on the sales floor? What about the trucks and trains to move items about? All are integral to the business. So from a scope standpoint we need to consider almost all of the content within this type of enterprise. The key guidance to apply is—the scope of DG is a function of the assets being managed (i.e., the content and information being governed).

Federation—the entire enterprise is in scope, and all content relevant to the business model is in scope. We have not narrowed this down much, have we? When we examine the content (remember we are considering all of it), we see that it stratifies into global, regional, and local. This is significant. If a region or locality can buy items to sell, what is the intensity of DG in those supply chains versus the global ones? We have to consider that local data may not be worth close governance and may be okay with a more relaxed level of intensity.

Bottom line: All content is in scope, but due to size, geography, and markets, we need to consciously identify which specific content is managed centrally, regionally, or locally. The organization would state that DG scope is all content relevant to the business model, but the intensity of DG will vary based on a specifically defined set of federated layers.

Elements of data governance programs

In many ways, a DG program is like any other business program. Many elements of DG make perfect sense to businesspersons when they first consider DG. For some reason, the people on the technology side of the information management and DG equation get dazed and confused. Either way, this section will introduce these basic program elements in the context of DG. We then will review the important, managerial aspects of these elements. Later on, we will get into the specific design and deployment of each of these elements.

DG is simple in its makeup. There are the three cornerstones of any organization—People, Process, and Technology. On top of that, we need to add the newest, permanent cornerstone and aspect of any program—Data. (Do not stop at the People, Process, Technology mantra—you MUST add data).

The People-aspect deals with the various required roles. Process needs to cover an operating model and the capabilities required to operate DG. Technology covers DG-specific technologies. The Data element covers the actual content, metadata, catalogs, all content, data dictionaries, etc.

Because we are answering the "What does it look like?" question, let's briefly cover the Process element first.

Process—The DG operating model

The DG operating model is made up of two components: the capabilities model, which states WHAT is happening, and the workflow model, which dictates how information flows and parties interact within the program, and how decisions are made.

Essential capabilities

A capabilities approach has become more popular for addressing DG functionality over the years. It is familiar. The basic approach to standing up DG is handled like standing up any other new business capability or upgrading current capabilities to include better data behaviors. We use the term capability to describe the "what" has to happen in DG. Because it is programmatic, DG introduces new capabilities to the organization.

Capabilities are also much more stable than processes or functions. Since DG is a long-term alteration in data operations, it makes sense to be more stable. In a world where agility is important, capability-based design allows iterations more easily, since your target is putting in stable capabilities vs a specific process tied to a policy.

Initially these functions will appear to be embedded in the DG "area" but over time they need to evolve into day-to-day activities within all areas. Using the word "process" would immediately imply a "where," and that is an operational level of detail that evolves. Within the capabilities are functions and activities that will be visible in the DG management framework as stand-alone processes as well as the day-to-day activities that carry out governance. There is no need to design your DG functional model from scratch; there is a list later in the book. However, recognizing that there will be a formal set of functional requirements (to be manifested as processes) and that they will be executed all the time is a key element to the success of DG. Capabilities perform two roles. First, they point out what someone actually must do. Second, reviewing the functions required for your organization usually aids in determining which areas or individuals would bear accountability and responsibility.

The DG area will need to consider other business areas where there will be interaction and collaboration, such as:

- Human resources
- Compliance and/or legal
- Risk management
- Large-scale integration projects, such as enterprise resource planning

The bottom line for this element of DG is that you need to formally consider and build the DG processes and functions. They are not instinctive.

In the first edition I called out FUNCTIONS to indicate the WHAT. However, "Capabilities" fits better. Lastly, this is the language of enterprise architecture, and over the past years I have noticed it is easier to get DG engaged with enterprise architects, as well as the usual constituents. Table 4.1 shows a very high-level representation of DG capabilities.

Table 4.1 High-level data governance capabilities
High level data governance capabilities
Data and governance strategy
Data governance requirements and design
Data governance frameworks
Supporting technology
Data governance operation
Data governance engagement
Data governance measurements
Technology operation
Communication
Training
Data governance services
Sustaining design
Sustaining activity management
Sustaining program operation

Work flow

Like any other activity within a company or government entity, there needs to be a formal statement of work flow. Please note that this is no hint to design an organization chart. Rarely does DG activity become a stand-alone area (i.e., there is rarely a DG "department").

Work will flow within DG like other forms of governance. In lieu of an organization chart, consider DG structure to be more of a matrix. This allows incredible flexibility (Table 4.2).

Table 4.2 Data governance as work flow		
	Management	**Operations**
Strategic activity	Set strategy, align to business, provide principles, ultimate issue resolution	Specify controls, metrics
Tactical activity	Identify issues, offer coaching, support application of policy	Provide DG services, operate supporting technology, provide training and communications

Perhaps the best way to see DG capabilities in action is to look at a common artifact from an active DG program—a meeting agenda for the main DG council. The council represents a group that gets the governing done (Fig. 4.3).

Council Agenda Planning				March 1, 20nn
Month: **March 7th**		**CURRENT AGENDA**		
Presenter	Topic	Item	Duration	Action required
	Orientation	Charter ratification discussion	10	Understand role of the Data Governance Council (DGC)
		Principles and policy introduction	20	Understand role of principles and policies
	DG Value	Recommended DG metrics	5	Understand role of metrics to measure DG and review proposed measures for DG
		DG scorecard introduction	10	Review current DG scorecard
	DG Compliance	Principles and policy roll-out	15	Determine / approve principle / policy roll out
		Status of classification policy and efforts		Awareness of effort and business impacts
	DG Process	DGC meeting protocol	10	Define meeting protocol
	Issues	Custodian training	20	Trial issue—Review and recommend resolution
			90	
Month: **April 3rd**		**FUTURE AGENDA**		
Presenter	Topic	Item	Duration	Action required
	DG Process	DGC meeting protocol, rules of order	10	Define meeting protocol
	Orientation	Education—Role of DG	10	Undertand role of principles and policies
		Demystifying policy presentation	10	Undertand policy hierarchy
		Sample DG scenarios currently underway	15	Undertand what is going on
	DG Roll-out	Crawl, walk, run approach	15	Review and adjust/approve DGC "road map"
	DG Value	Short term objectives—walk the talk	15	Leave with consistent identity and message
			65	

FIG. 4.3

Sample data governance agenda

Helpful hint

Remember that the ultimate goal of DG is to disappear as a stand-alone program. It becomes part of the fabric of business, like financial controls. That is why the DG "department" is really a monitoring structure—much like an audit committee. To be more specific, DG may not really disappear, but it will be very thin. There will always be the need to resolve issues. But like other types of corporate governance, these events become accepted as normal activities, not special programs.

There may be a thin department with participants that roll in and roll out. Some highly regulated organizations may want to have a separate DG department only if it cannot be fit into the compliance areas. This is one of those areas we promised to point out—we differ from many of our peers. Given the long-term nature of DG (i.e., it's not a lot different than financial controls or well-known policies), there is little need for a full-time overhead structure. In our opinion, that perpetuates the labeling of data management and governance as programs that can be terminated, as opposed to the permanent behavior change it really represents.

A specific time frame for DG to reach the "transparent" stage is hard to define, as it will vary based on scope and organization. The degree of transparency will be tied to your progress up the IMM curve, so whatever the timing is of your IMM progress is most likely your timing for DG to fade into the business fabric. If you are using a five-stage maturity model to measure DG effectiveness, then whenever you hit stage 4 or 5, your DG program should be part of your everyday activity. This may take a very long time. Think about it more as a goal rather than a requirement.

Again, think about financial controls. Few organizations talk about the financial governance program and whether it should be justified to continue or be terminated. That would be unheard of. This perception is what you are striving for with DG.

Regardless of the size of the organization or the complexity of DG, it is key to remember the DG organization is not there to do information management. There will be DG functions and processes to manifest the capabilities. (There is a complete list of sample functions in the appendices.)

Principles

We touched on principles earlier by way of a definition. In summary, they are general adopted statements of philosophy that guide conduct and application of data management and governance capabilities. Principles are more than just a term to be understood, however. Principles are crucial elements in DG. One client told me they justified the entire program because with principles in place, there were fewer meetings.

Principles will succeed where a batch of rules and policies will not. They are foundational. One explanation we use when confronted with resistance to developing principles is to draw an analogy to the "Bill of Rights," the first 10 amendments to the United States Constitution. It's easy to see the historical significance of the application of these principles to United States history. It is the same for data principles (maybe a bit less historical).

As you deploy DG, you will need to revisit and repeat your enterprise-level principles. Not revise but repeat. Since they are foundational and represent beliefs, repetition will be necessary. Table 4.3 lists some sample principles we have collected. (Please note that I occasionally refer to an organization named "Farfel." This is the name of a fictional company used as a sample case study and will appear again in later chapters.)

Table 4.3 Sample principles

Principle name	Principle description
Master principle	Enterprise data will be governed by a formal organization, with appropriate authority and accountability to define and establish how information, data, and content is managed.
Federation	FARFEL will have enterprise standards and guidelines for all metrics, content, data structures, codes, values, and data naming.
Data efficiency	Data, information, and content needs to be available at the right time, at the right place, and in the right format to authorized users/consumers, at an efficient cost.
Business alignment	Information management solutions will maintain business alignment, and will only be in response to business needs vs business area requests.
Information quality	Enterprise data will be managed and measured for quality. There will be parties that are accountable for overall integrity and quality of enterprise data and content.
Risk management	Appropriate due diligence will be conducted to ensure data complies with all applicable statutes and regulations.
Share and collaborate	Enterprise data is a shared resource across the enterprise. Data is not a resource that can be "owned" by specific business areas.

Policies

Another element we previously defined is policies. Policies are formally defined processes with strength of support—that is, they are a codification of a principle. They give it "teeth." Policies include standards—one area where IT personnel will be very intense as DG becomes real. Most likely, you already have most of your DG policies floating around in the form of a disconnected IT, data, or compliance policy. And, like most places, the policy sits happily in its notebook while life goes on and the policy is disregarded. The marriage of principle and policy prevents this in the DG program.

Metrics

You cannot manage what you do not measure. Over time, your DG program will need to evolve a means to monitor its own effectiveness. Without it, the DG program will certainly fade away. At the outset, the metrics will be hard to collect. After all, you have not been managing data very well, so there is no infrastructure to install a metric. Eventually, the metrics will evolve from simple surveys and counts to true monitoring of activity. By way of explanation, here is a brief list of common metrics:

- *DG Stewardship Progress*—Report on counts of individuals trained on DG, counts of specific projects governed, and a count of issues elevated and/or resolved.
- *DG Stewardship Effectiveness*—Alternatively to progress, an effective metric can be based on counts and resolution of issues submitted to DG bodies.
- *Data Quality*—Data profiling results calculated into a DQ index that represent an average of all of the data-quality profiling measures.
- *DG Value*—We will dive into the business case and business value more in the next chapter, but you can never go wrong with tying the application of DG and data management to business success. Quantifiable and intangible benefits resulting from successful efforts that were governed, or through use of governed and well-managed data, should always be reported.

The people element—Roles and responsibilities

The official designation of accountability and responsibility are key factors to the survival of DG. Most important to new DG programs is the concept of accountability for data. This is most likely a very new role. To be clear, it will seem very new and different to hold someone accountable for data quality—especially when accountability means a direct effect on bonuses or promotions. There will also be a perception that the DG program is rather powerful or bold to be making these designations. Assigning responsibility will also be an important activity. In many organizations, the responsible parties have a formal role as designated "stewards" or "custodians." Other implementations of DG may place everyone under a label of a steward, and the responsible parties will be direct supervisors.

Many organization view DG as a door through which to introduce new roles. The new roles tend to focus around a concept of placing an individual in charge of data assets, usually at a level above and outside of information technology. This "top data job" places formal accountability for management of data in one place. The Chief Data Officer, or CDO, is a manifestation of this job. Some CDO or equivalent titles will report to the CEO. Others may report to, or act as, an office in charge of data monetization or analytics. Either way, these organization find that data assets are important enough to warrant a new line of authority. In my practice I look for opportunities to get a top data job identified and placed. It helps a great deal.

Data

Governing data means understanding your data. That means the operation of DG will help oversee managing data wherever it is, and however it moves around. When planning and operating DG, this means the program needs to be aware of the data landscape.

The data landscape is your inventory of data—what do you have, where is it, who uses it, where does it come from. Often a lineage or provenance capability is also required as part of managing the data landscape.

The data assets of the organization, represented by the landscape, are the operational targets of DG, whereby it oversees the usage, movement, interfacing, and integrity of the data.

Most issues in DG will come from application of governance to a data area that is confused or resistant to change. Hence, Data needs to be a distinct focus area alongside People, Process, and Technology.

Technology and tools

The last element that requires high-level consideration is Technology. As of the writing of this edition, the market for pure DG technology is evolving rapidly. Specific tools for data glossary management, DG workflow, data discovery, and data quality/governance integration are being brought to market. These are on top of the traditional technologies in support of DG, such as SharePoint, Word, and Excel, as well as adapting tools from other disciplines, like data model or data dictionary tools. Specialty tools are evolving and, in general, you will want to consider the following capabilities, but Chapter 15 will cover the application of tools in more detail.

One aspect of tools to understand at this point is that you should not feel compelled to buy DG tools just because you are doing DG. By definition, a tool exists to improve something you are already doing. If you are not doing formal DG yet, or if you are doing it poorly, then casting about for a tool to help you deploy DG is a waste of time. This flies in the face of typical IT philosophy, where the tool is usually acquired first. It's a notoriously silly thing to do. Our work always has us putting the brakes on a tool selection project: It's easy to buy a tool and install it. But most of the time we witness new tools for data management sitting unused or poorly deployed, because no one has mastered the process the tool is supporting.

As you roll out DG and begin to understand the various aspects of your particular program, you will know immediately where you need a tool to "grease the skids." Some features of DG tools to consider are:

- Principle and policy administration
- Business rules and standards administration
- Organization management
- Work flow for issues and audits
- Data discovery
- Taxonomy or ontology management
- Data dictionary
- Enterprise search
- Document management
- Metrics scorecard—data gathering, synthesis, and presentation
- Interfaces to other workflows and methodologies
- Training and collaboration facilities

The critical success factors for data governance

Because DG is a business program, we need to point out the critical success factors (CSFs) early on in this book. Frankly, if one or more of the CSFs presented next are totally unrealistic for your organization, you need to reconsider launching a formal DG program as an approach to improving data asset management. Or at least, you should call it something else.

1. DG is mandatory for the successful implementation of any project or initiative that uses information. Any project requiring reports, business intelligence, cleaning of data, or development of a "single source of truth" requires DG to be sustainable and successful.
2. DG must show value explicitly. This means you cannot do DG in a vacuum. Something has to be governed, even if it is data quality and you implement DG as a means to improve data quality. Countless IT shops developed models, standards, and policies in the 1980s and 1990s, and then went looking for a project to spring them on. You need to show benefit, and that means tying the DG effort to a visible initiative.
3. You must manage organizational change. At the risk of being repetitive, you are doing DG because you are NOT doing something correctly. Therefore, something needs to be changed. We have dealt with numerous organizations who wanted all of their data fixed but did not want to change their views or the behaviors or processes that created the mess. So, you will need to orient, train, educate, communicate, hold hands, encourage, and offer incentives. Then repeat it all again.
4. DG must be viewed as an enterprise effort. You can implement it in segments, but it must always have an enterprise perspective. Otherwise, you will end up with conflicting standards and accountabilities.

What about Gladys?

Far too often management sets formal organization change management aside. Usually the reasons are [with a valid response in brackets]:

- We don't have enough time. [It does not take very long...sorry.]
- We cannot afford it. [DG has a net cash cost of zero. Plus, can you can afford to have the project crater?]
- It is squishy. [Anything that, if done improperly, can cost you millions of dollars/euros/pounds, etc. is not squishy; and there is data behind that statement.]

But, in spite of piles of data proving this is not a very bright thing to do, it still happens. If you are experiencing this problem, consider (or use) this story:

Gladys works in procurement plant in Iowa. Every day (for the last 20 years), she logs on to four applications to do her job. Once a week, she downloads operational data into a spreadsheet and prints out the weekly inventory updates for her boss. From a data standpoint, she is the sole integration point for three operational systems affecting finance, work orders, and inventory. She is proud of her accumulated knowledge that allows her to accomplish her duties, despite the poor data management aspects.

When the company finally fixes the kludge of applications, should the new processes and training be handled via:

(a) An e-mail on Friday that comes Monday; she has a new password and the instructions will be on her desk?

(b) A change program that has her participate over time to define the new interfaces and processes, including flying her into headquarters to meet others going through the same process?

(c) Having the vendor of the new software stop by and do a one-day training class?

Obviously, (b) is the kinder method, but (b) stems from the change management discipline. Sadly, (a) and (c) are the more typical approaches because no one ever gets to consider (b) since the organizational change effort was squashed.

The human issue is critical—What if Gladys was your mother?

Summary

Businesses are accustomed to controls. All organizations have a standard means of ensuring the integrity of financial assets. There is not a single CEO on the planet who would condone multiple sets of accounting standards in their departments. DG is no different.

The DG program offers a set of capabilities that behave like any other business program. Most of the time, DG is a defined operating model that sets out how decision makers operate. It is a framework for better behavior to enable issue resolution, monitoring, and direction setting. It should NEVER be thought of as a set of new processes to be accommodated.

It is not easy, but the entire enterprise needs to accept that twenty-first-century organizations' dependence on data assets implies the acceptance and institutionalizing of a DG program.

Essential questions

1. The main, typical components of a program are People, Process, Technology, and Data. How are these applied to a DG program?
2. Why are Principles so important?
3. If your organization had offices in 20 countries, what would you consider when scoping your DG program?
4. What is the long-term goal of a DG program?
5. Why is the assimilation of DG into everyday organization life difficult?

The data governance business case

5

Chapter Outline

A manager is responsible for the application and performance of knowledge.
Peter Drucker

As stated before, data governance (DG) is a business program. Most successful DG programs are actually sponsored by non-IT areas. This has stemmed from a realization that a large percentage of the money spent to date on various data-related, IT-managed programs have not met expectations. However, since DG is a program that seems to deal in abstracts (data as an asset), it is similar to other programs where tangible results are hard to see, such as marketing. The CEO will acknowledge the need for marketing and certainly the need for a finance area, but a detailed, hard-dollar justification for these areas (as for DG) is usually not sitting in a folder on a desk somewhere. That does not excuse any DG effort from presenting a business case, however.

Data Governance. https://doi.org/10.1016/B978-0-12-815831-9.00005-9

In spite of DG being a new capability with a goal of minimal cost increase, leadership is owed a statement of value, and a return on whatever investment is made.

The business case

The logic is simple. DG is a business program; therefore, it needs to add value to the business. Any long-term cost incurred that exceeds benefit or risk reduction is unacceptable.

However, obvious, tangible results are hard to see. The DG program will often have a weak business case. Far too often I hear that DG is required to "improve report accuracy" or "ensure better decisions" and many other lightweight objectives. These cannot take the place of a specific identification of business benefits. Leadership may receive some criticism for remaining data illiterate, but many DG groups get equal attention by creating insipid business cases.

Can you be excused from a business case if the CEO says, "I know we really need this, and it is like marketing—so proceed without a business case."?

No.

The treatment of information and data as an asset should tightly connect data to business activity. *A business case is required even if it is not requested.* There are several reasons for this:

- DG is a holistic effort requiring enterprise attention at some point (even if you start with a low profile). There will be naysayers and you need to be able to handle them. A common form of resistance is for a department head to state there is no time to participate on a new committee or learn new procedures. After all, there is a business to be run. However, it becomes harder to throw resistance up in the face of a business case tied to a goal of making hundreds of millions of dollars for the organization, or better yet addressing the very issues that the department head is struggling with.
- DG will not succeed if it cannot be measured, and the success measures must come from a set of business-oriented metrics. Metrics around "better decisions," or "timely reporting" do not count (more on that when we cover metrics in a later chapter).
- DG tied to a specific project does provide a sustainable business case. There may be overwhelming data-quality issues or strong pressure from regulators. A data lake may not be meeting expectations, or analytical models are failing due to data quality. There may be a large implementation of an enterprise resource planning (ERP) package planned. DG becomes a necessary part of these projects and will make a direct contribution to success. The ERP has its governance needs, the data lake its governance needs, etc. All of these scenarios create a risk of developing sets of similar yet non-united DG activity. The issue is you cannot attain a sustainable program this way. *Without the enterprise business case DG is "dumbed down" from a business program to a business interest that is then passed to IT where it becomes a project.* This progression, of course, directly conflicts with the essential aspect that DG is an enterprise effort.
- The insistence in many organizations on developing a hard-and-fast business case with "real" benefits and strong financial returns that are based on traditional benefits (like headcount reduction or reduced business costs) is an obstacle. The business case for DG is marginalized, or DG is ignored because it cannot do what is viewed as a "real" business case. Financial management believes a business case with tangible returns is impossible because managed data and content are "intangible." And the business case is deemphasized, or we manufacture faux benefits based on technology efficiencies. So a bit of creativity is required to convince that this very narrow thinking can be converted into a legitimate business case with "hard" benefits.

Objectives of the business case for data governance

Showing the value of DG is accomplished in two ways. First, the value is shown in the form of a tangible direct benefit, where you can tie DG to benefits coming from one of four directions:

- Increase in direct business contributors, like revenue, customers, or market share (e.g., postmerger economies of scale, efficient supply chains, effective promotions)
- Improvement in efficiency (e.g., integration, faster information delivery, enabling or empowering employees)
- Monetization of data, in the form of selling intellectual property, or creating new products with new features derived from or including data. (e.g., selling depersonalized business activity to third parties)
- Reduction in risk, either through fewer fines, lower reserves, loss of market share or reduced cost of risk management, such as insurance premiums (e.g., compliance to General Data Protection Regulation (GDPR), improved information privacy, improved data quality). Also avoiding the risk of reputational issues through breaches and ethical use of data. In many organizations, the easiest direct benefit is derived from reduction in risk. Three or four decades' worth of explosive growth of stored data and documents has created enormous amounts of risk. A few examples of this are:
  - Privacy violation
  - Data security
  - Civil liability brought on by poor management of safety or warranty information
  - Incorrect decisions brought about by inaccurate or inconsistent data across numerous copies (e.g., establishing reserves too low, or losing track of where you acquire items)
  - Regulatory liability by failing to track key documents or respond to a request for documents
  - Unethical use of private data
  - Excessive costs keeping ROT (redundant, obsolete, and trivial) data, including documents, backups, SharePoint, and e-mail

The second form of tangible value is indirect, in much the same way as a marketing program (i.e., the marketing program will support other initiatives that would otherwise fail or falter without the program). There is also an indirect form of value in reduction of future obligations to deal with bad data.

In the case of marketing, value is determined by predicting and confirming increased market share or more prospects. Marketing strives to improve visibility of a product that, for example, supports more sales. In a similar manner, the value of information projects stems from where the information is used. Therefore, the DG business case needs to support the activity that ensures good data and information is available to accomplish business goals—without incurring undue risk or cost.

You need to look for opportunities where DG supports business programs that want to increase revenue, lower costs, and reduce risk. Once you have identified opportunities to aid in achieving business targets, then it is time to specifically quantify business benefits and align them, in detail, with the data and content that DG will be overseeing.

In the case of dealing with the future cost of bad data, you need to strongly consider a business case that incudes management of "data debt." When an organization chooses to defer doing the correct behavior with data assets, whether intentional or not, then it incurs a cost in the future to correct the misstep. Like any other debt, it needs to be paid or written off. A business case for DG will make a clear statement that can often be quantified, concerning the future cost of not taking the right action now with data. A common scenario that demonstrates this occurs when an organization is presented with the need to develop a source of data to solve a problem, such as an external regulatory report. The data most likely exists in various locations already but requires some hearty maintenance to make it usable if it stays within the current databases. Or the temptation exists to scramble, manually patch up the data, and create (yet another) a stand-alone database for the report. Ms. Executive says, "go ahead—I need the report. I do not care that we have another database." And that is true *unless* you point out the future burden that has been created.

Another objective of the DG business case is to build a response to the historical shortcomings of IT and unsupervised data projects. These are:

- The perception that data and information initiatives always fail
- The perception that spending on "pure" information management projects is wasteful
- Development of large "lakes" or "warehouses" of data, that are understood by only a few data scientists and analysts, resulting in expensive bundles of technology benefiting only a few business areas
- Ongoing complaints that the IT data is not "correct"—so business areas need to create "correct" data
- A growth of "stealth" or shadow IT in reaction to a poor perception of IT
- Lists of projects that "we will get running with these shortcomings and then fix them later." Of course, later never happens.

The DG business case must address these opinions head-on. To recap, it needs to accomplish the following:

- Identify where it can support business directly (such as risk avoidance).
- Identify where data and information is used to move the business forward requiring the enabling capabilities of DG.
- Associate DG with data management capabilities (master data management [MDM], business intelligence [BI], data quality, data integration, data movement, etc.).

Accomplishing objectives like these will provide a multidimensional business case that will make DG a sustainable program.

If detailed, specific business benefits cannot be quantified easily; you can use industry standards, benchmarks, and papers to provide the metrics for the business case.

Components of the business case

Several basic elements are required to build a business case for DG. Because DG is a component of the overall enterprise attitude toward data, there are similarities in the DG and data management business case and other business cases. But there are also some differences. A lot of other details on enterprise data business cases can be found in *Making EIM Work for Business* (John Ladley. Waltham, MA: Morgan Kaufmann, 2010). The basic contents are slightly modified for a specific DG case.[1]

---

[1]That wasn't shameless promotion. There are a lot of templates and information for data management business cases in that book. Borrow it from a friend.

The big picture (vision)

Vision is perhaps the most abused term in business, but the "big picture" is incredibly important for the acceptance of DG. Remember that you will be requiring a large part of the organization to change. Change does not happen among humans without some view of the big picture. In fact, it is rude to ask people to change without some sort of explanation.[2] This is your goal for the vision. *What will a day in the life look like when DG is in place? What will you see in the organization? What business goals will be more achievable?*

One of the big surprises in rolling out DG occurs when the business areas start to comprehend that there will be new accountability for data. Very often an oxymoron will develop. The same business units that insist on their own IT staff and maintain scores of legacy spreadsheets and Access databases will also say, "Data accountability is not my issue. Data belongs to IT. Except my data, that is."

Never say "better decisions" or "better data quality" as business vision statements. These are not business statements. They have no relevance from a vision standpoint because they are not measureable in terms of business value, and they improperly position expectations. An example of a properly worded business vision for DG might look like this: "ACME, Inc. will manage its information assets to increase shareholder value and reduce enterprise risk."

Program risks

While the business case is a vehicle to present how an enterprise will manage its risks, you also need to consider the risks that the DG program itself may create:

1. *Business Risks*—The DG program fails to do its part to prevent loss of market share and reputation and fails to hit targets or avoid fraud.
2. *Regulatory Risks*—DG fails to address compliance requirements and there are regulatory violations.
3. *Cultural Risks*—The organization fails to engage in the DG process and continues the poor data asset management practices that resulted in the need for governance in the first place.

Business alignment

If the DG program is going to be supporting (directly or indirectly) business initiatives, call out the value points or specific scenarios enabled by DG. Your actual business case benefits will come out of these areas, so do not be timid in looking around for opportunity.

Costs of data quality issues

Data quality issues consume an enormous amount of cost and resources. It is the primary manifestation and metric of a functional DG program. Therefore, it is important that your business case mention the current costs and risks associated with data quality.

---

[2]When raising children, parents take great pains to never say, "Because I said so!" as a justification for a desired behavior change. We all know that is not easy. In the context of DG, the temptation to say "Because I said so" with difficult cultures will be much stronger.

Costs of missed opportunities

There is always the need to highlight what will happen, or continue to happen, without DG. You may cover some of this in the data quality area, but it is good to recap existing issues with data, reporting, poor content management, scary compliance issues, or the high cost of ownership due to extensive redundancy. In addition, there may be business actions and scenarios that cannot happen or may be more difficult without DG.

Data debt

If there is an awareness that DG will be reducing accumulated data debt (all organizations have some), it is a good idea to mention the concept and try and calculate what has been accumulated. This is the amount of money it will require to fix the data problems—it is estimating a project that there is usually no intention of funding. Alternatively, you could look at all of the current data issues, and report on a rough estimate to fix them all. Until the debt is paid, you will always pay more to maintain your data landscape than you should be paying.

Obstacles, impacts, and changes

It is fair to cover possible cultural and other organizational issues. If there is the possibility of technology changes, these can be mentioned (you do not need details, those come later). Any obstacles that are known need to be presented.

Presentation of the case

The business case for DG is a business document. Even if the CIO is handling this task, you need to avoid three-letter acronyms, techno-babble, and exotic and abstract pictures. You are selling—and any salesperson will tell you that you must be crystal clear and concise.

> **Helpful hint**
>
> Do not depend on a single presentation to sell the DG program. You should be vetting ideas and benefits long before the final PowerPoint blast. Know your audience (i.e., who will be nodding yes, shaking no, or nodding off) before you even schedule the final presentation. The best scenario for a final business-case presentation is a 30-minute review with key decision makers and their acknowledgement that everyone is okay to move ahead.

A few themes must dominate the business case:

- DG is a program. (Even if the ultimate goal of DG is to become woven into the enterprise, it is still programmatic in its rollout and lifespan.) You are funding a long-term, permanent change in mind-set and behavior, but the organization won't embark on this journey without some form of return or perceived benefit.
- DG is supportive of many projects but, most importantly, it is the control and audit function for managing data assets.
- Governance and change are mandatory to address the issues that created the need for this meeting. Make sure those issues and history are understood.

At the highest level, a short and concise presentation is required. My guideline for a CEO-level briefing is ten slides or less. If the presentation is done well, the DG team should expect an expression of interest, commitment to proceed, and feedback. The CEO's feedback must be an acknowledgment or correction of the business alignment items and must convey an understanding of the risks and impacts. If this material is presented to those at lower levels in an organization, then add details around impacts, business benefits, and risks.

The process to build the business case

What follows is a brief outline of the process to develop the business case for DG.

Fully understand business direction

Whether you have explicit access to corporate strategy or need to read the annual report, you must form the DG business case in the context of your organization. That means not accepting a boilerplate justification from a conference brochure. Why is DG relevant to *your* business? If you are forming DG as part of a broader enterprise information management (EIM) effort via MDM, BI, or both, then confirm that the DG team knows where the business wants to go.

Identify possible opportunities

Business strategy begets information opportunities. Again, if an EIM program is being implemented, you may have this information handy. A common type of direct benefit of governance is in the areas of e-discovery and document management. It is where organizations drastically reduce cost and risk of document handling by simply implementing better governance.

Identify usage opportunities

The indirect benefits of DG come from efforts where information is used to deliver a business result, such as a data warehouse. In these cases, DG can help ensure a consistent and relevant result. If there is a large customer MDM effort tied to some sort of customer program, then your DG effort supplies the required governance to the new MDM policies, standards, and processes.

Define business benefits

Refine the potential benefits in terms of not only a perceived high-level number, but also in terms of cash flow or earnings increase. In addition, describe specific risks. Look for risk across the three risk types—regulatory, civil, and financial.

Confirm business benefits

Confirm business benefits you have identified to ensure they are supported by DG. Make sure you do not attempt to support something that is not relevant.

Quantify costs

Examine current costs of IT as well as other information-related costs, such as the numerous departmental business analyst, databases, and tools and external data sets that did not go through IT oversight before being acquired. Include all capital costs, depreciation, and overhead. Any analysis of the cost of poor data quality should be factored in here as well. Include costs of departmental end-use databases, spreadsheets, and "ShadowIT." This is a good beginning cost number. It points out how much is being spent now, without governance. The actual cost of governance should be a small fraction of current costs. Ideally, you will use internal resources. Most of the time we initially see a small increase in costs for some consultants or for training, but as DG becomes part of the enterprise, costs decrease or return to prior levels.

Prepare the business case documentation

Apply the various financial benefits and costs to whatever model is used or selected by your organization; then present the results in whatever format is palatable.

Approach considerations

Many, if not most, companies do a horrible job disseminating their business plans, and that assumes they actually have one. I have been involved with dozens of data management and governance engagements over the past 30 years. Few of these organizations had a business vision or strategy that was readily available to the very people whose job it was to ensure those plans could be measured. Often the request for a business strategy triggered an embarrassed fumbling in a cabinet during an interview: either a plan would be produced, or there would come a "need to know" denial. Organizations that do publicize their strategies and push this information to all levels tend to have much less challenging information and content needs. This is not a coincidence. If business drivers and goals are endemic, how hard is it really to match up the applications portfolio and business intelligence efforts with the business direction?

Helpful hint

Business alignment

Any formal business alignment exercise will demonstrate how business and information/content usage is connected. This is what positions the organization for a formal business case. It means taking any business alignment material that you have already prepared and starting to use it.

It is at this point that organizations that have not done business alignment stop, hire a consultant, and then do an alignment exercise. Let us then reinforce the importance of business alignment—it will be done regardless. The issue is to do it early on and in full understanding of the relationship of EIM as a business program within your enterprise.

In the typical scenario the business plan has been developed, but it is considered "top secret." This can be early resistance, lack of engagement, or a form of top-down misdirection. Obviously, you can have secret strategies and still give middle and lower management enough to discern business alignment. It is already in their performance objectives, isn't it?

The plain and simple fact is this—if everyone knew where the business was headed, many of the information management issues we have covered would be minimized or eliminated.

Summary

Even if a business leader clearly trumpets the need for "better data," and is willing to push hard and use political capital to get it, you do not go forth without a business case. If you do, you run the risk of falling into the garbage can of failed initiatives. Thus, there are some business considerations for the business case as well:

1. The business case must address accountability. If the goals are not met, who is responsible? Historically, it has been very easy to blame IT for a failure to communicate. A clear business case will use business terminology and point out where the business accountability is.
2. Business leaders are poorly incented to do well at information-type projects. The business case for DG must support business accountability and be built into the sponsors' objectives and personal targets.
3. Once IT projects "happen," there is a tendency for interest to wane, and even return to the old alternative. Business areas need to understand that the investment continues beyond deployment, and some effort and willpower are required to sustain the project's goals. The business case must acknowledge the cultural impact and even accommodate the costs and benefits of sustaining the effort while ensuring changes are fully adopted and integrated into the fabric of the culture.

The DG team needs to remember that there must be a sales process of sorts, even if none is requested. A proclamation from the CEO, or the best possible sponsor does not make a successful program. This means examining business opportunities, educating about the ramifications of managing information as an asset, and recognizing that the long-term animosity between IT and business areas must be addressed with a business program. Don't forget there are challengers and naysayers out there. Starting a business case tied to clear opportunity with a solid financial impact will help slow down early resistance. Fig. 5.1 shows the summarized, initial business case for the Rocky Health case study which will be presented in detail later. Like all the artifacts in this book, it is an amalgam of several real examples. It is hard to offer blatant resistant when you realize that without DG, these benefits are not fully achievable.

Data governance controls data usage in order to achieve:			
Tangible benefits—known and documented amounts			
Reduce risk	Reduce average level of fines and holdbacks by 80%	Average $3 million past two years	$ 3 million
Achieve operating margin	Reduce nonvalue added efforts	Addressing fines and errors accounted for 17% of gross operating margin	$7.5 million
		Hired 16 BAs across 8 departments at. 4% of gross revenue	$1.1 million
Wellness—Improved patient outcomes	Balance populations served	Attracting 1% more non medicare patients	$2.3 million
Intangible benefits—known financial and reputation impact but hard to quantify			
Data landscape needs to become cost/risk aware	Stop buying multiple tools	Physicians purchased 4 cloud-based reporting tools w/o IT awareness last year	$50,000 / year subscriptions PLUS time wasted over arguing whose data is correct
Greater focus on action vs. reaction	Mid-level managers are crisis oriented	Duplicate efforts to solve similar problems	As much as 8 FTE worth of duplicate efforts
Realization that current methods add enormous overhead	FY2017 budget not approved until June 2017	Most of delay was as a result of in accurate project and labor data	Overspending—hard to quantify but as high as $10 million
Clearer communications	Four areas are "in charge" of some sort of reporting or analysis	Duplicate efforts to solve similar problems	Duplicate FTEs plus frustration

FIG. 5.1

Sample data governance business case

Essential questions

1. Is a business case always necessary for a DG program?
2. What are the objectives of a business case for DG?
3. Can you think of a strategic initiative that does not require, affect, or use data?
4. What is data debt?
5. Would data debt discussions be useful in your organization?

Overview of data governance development and deployment

6

Chapter Outline

> *If you don't know where you are going, you'll end up somewhere else.*
> **Yogi Berra**

This chapter is the last chapter where management and leadership are the target audience. It is also the first where we start to get into the "how to." It is also very different from the first edition. Data governance (DG) practitioners have been learning a lot since the prior edition, and the attempt is made here to convey as many lessons learned as possible within the constraints of this book.

This chapter is not a mere enhancement of the first edition. Experience, technology, and pace of adoption have caused the approaches to DG to evolve. Franky, most of the tasks are the same. However, their arrangement and context have evolved. The framework we will review offers greater agility and flexibility.

This chapter lays out activities at a high level. While reading you may have the impression of a linear process; that is not the intent. There are many ways to get from a current state to a future governed state. You can be low profile, or very direct and controlling. The process you use can be linear, iterative, or agile. The areas of activity covered in this chapter represent, again, your checklist. The activities can be blended based on levels of detail based on your own needs.

Data Governance. https://doi.org/10.1016/B978-0-12-815831-9.00006-0

The various tasks and deliverables are presented as a delivery framework, that can be adjusted to culture, priority, scope, and budget.[1] Your approach depends entirely on what is going to work in your situation, and we will cover the process to determine what will work for you as well as all of the potential activities that can be done to stand up DG as a new enterprise capability.

By "stand up," we mean establish engagement, start a program, define, design, deploy, and start to operate the DG program. DG is not a program to do as a stand-alone effort. After all, you need to govern *something*. So, the framework will also cover engaging with other areas. What is not addressed is a step to determine if DG is needed or not. It is assumed that an organization that acknowledges something needs to be different with its data will then get started and do DG. We also assume the intent to create capabilities with a visible benefit to the organization. We will not separately discuss if DG is a relevant discipline. If you have read this far, you know that.

This chapter will first present the considerations to determine your approach. Then we will break down the delivery framework, with examples of different approaches. The following five chapters will discuss a major topic in the framework with some sample artifacts and relate the steps to our case studies.

If you are considering DG, it means you have acknowledged a problem manifested through lack of governance. Therefore, one path to standing up DG is from the application of a solution to a problem. However, you need to keep the following in mind *all of the time*: DG is a component of an overall enterprise information management (EIM) program. As covered in Chapter 3, DG is applied when various types of EIM solutions are developed, such as business intelligence (BI) or master data management (MDM). Even if you are only doing an MDM solution and have no formal EIM program, in effect you are implementing one component of EIM. Since MDM and DG must go hand in hand, your MDM project lays a foundation for expansion of EIM through the DG and MDM efforts.

Helpful hint

We always emphasize that the "E" stands for enterprise when we talk about EIM programs. This is in reaction to the tendency of upper management to say, "First show me it works on a small scale." The same goes for DG. "Govern a little bit" is often heard during the initial days of a DG program. You need to be very careful that the understanding is once the "proof" (i.e., a business case) is shown, there is acceptance that DG is designed and deployed in the context of an enterprise, not a business area.

Lastly, DG can stem from concerns originating from a specific set of content. Strangely, while structured or "row-and-column" content is the first target of DG, many companies find themselves building fine DG programs when they clean up and manage documents. Databases and "row-and-column" data sets are governed long after nonstructured content, or a regulatory surge makes a company focus on a specific subject area. This happened in 2009–10 as fallout from the mortgage crisis and recession. Suddenly, DG became "hip." So, some of our activities ahead will address content of less structure.

---

[1] At this moment many readers are looking at the lovely flow diagrams from the first edition. No need to recycle them. They are still useful. In fact you can leverage both editions. However, one large lesson learned over the years is that it is better for the process to implement DG seem as flexible and iterative as possible. The first edition presented iterative processes, but no matter how many times it was mentioned came off as monolithic vs iterative.

Types of approaches

Consider the case study introduced earlier. You might remember that Rocky Health took a low-key approach. The initial activities had a narrow focus. The scope, of course, is enterprise wide, but initial implementation was kept focused. Now consider a global organization with significant exposure to international data privacy regulations, such as General Data Protection Regulation (GDPR).[2] There was an iron clad deadline and any organization that had to comply with GDPR has a lot of data-related work to catch up on. There was no low-key DG at these organizations.

The same can happen with a large applications project, such as a new enterprise resource planning (ERP) system. Some sort of global data oversight will be required regardless of the cultural ability to assimilate new data behaviors. So, as a best practice, the new data behaviors are part and parcel of an ERP training program.[3] Minimally invasive approaches work fine if there are no huge initiatives around. But efforts like ERP, being highly regulated, or data monetization means you have a broad focus from the start.

Your approach will tend (not always) to therefore move from very bottom up or organic, to more command driven, depending on the focus of the efforts *being supported* by DG. Repeating for emphasis: not the DG effort—the efforts being supported by DG. Standing up DG will be different when supporting a local, or low-profile effort, vs an effort with high visibility.

Fig. 6.1 shows some examples.

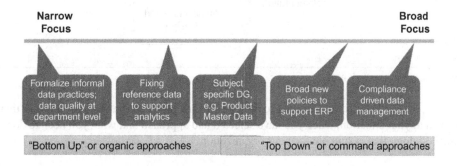

FIG. 6.1

Spectrum of DG approaches

Once you have understood the focus, your scoping efforts will allow you to select the best sets of activities from the development framework and put an efficient plan together. The approach also assists in determining the necessary DG activities to get started.

---

[2]GDPR explanation.

[3]Yes, this is a best practice. No, large integrators rarely apply DG concepts to training for a new ERP roll out. Too often within 6 months of roll out data quality is as bad as before the company spent $50 million on ERP. Sad, but true.

The data governance delivery framework

The delivery framework has five distinct types, or areas of work. Each area represents a collection of activities that can be used to further your DG program. Fig. 6.2 shows the framework as five phases, but only from the level where I need to show you all of the areas of activity in DG deployment and operations. Also, the efforts for starting and sustaining DG is shown as a cycle because it is usually iterative. Obviously, we need to show what the big picture looks like, but you will execute all or part of this cycle several times. *Additionally, you will rarely use the activities in each area the same way every time.*

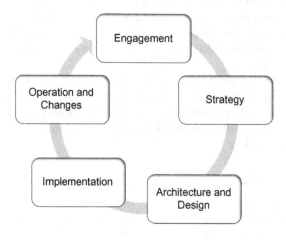

FIG. 6.2

Data governance delivery work areas

Therefore, the work areas we are going to review are not a recipe, but rather a framework that needs to be adapted to your situation. Once you have selected and approach and what you need to do you will have a methodology that is suitable for your organization.[4]

Process overview

This and subsequent chapters will be fairly detailed. In this chapter, the areas are listed along with key considerations. We provide a list of key outcomes and an example to provide context. The following chapters delve into the details of the activities and look at specific cases and deliverables. Fig. 6.3 shows the themes of each of these areas. As you can see, they can be stacked in a linear fashion if you require a large effort. But they also offer the ability to "mix and match" an approach unique to your organization.

---

[4]I am always asked at conferences or by clients to provide "how to" advice. Admittedly, it seems many consultants tend to provide "box and arrows" solutions and seem to be scarce when the hard questions are being asked. To be fair, you cannot understand the detailed steps unless you get a good dose of "box and arrows" learning. So, the earlier chapters were the framing chapters. Now it is time for the details. Remember, you asked for this. You have been warned.

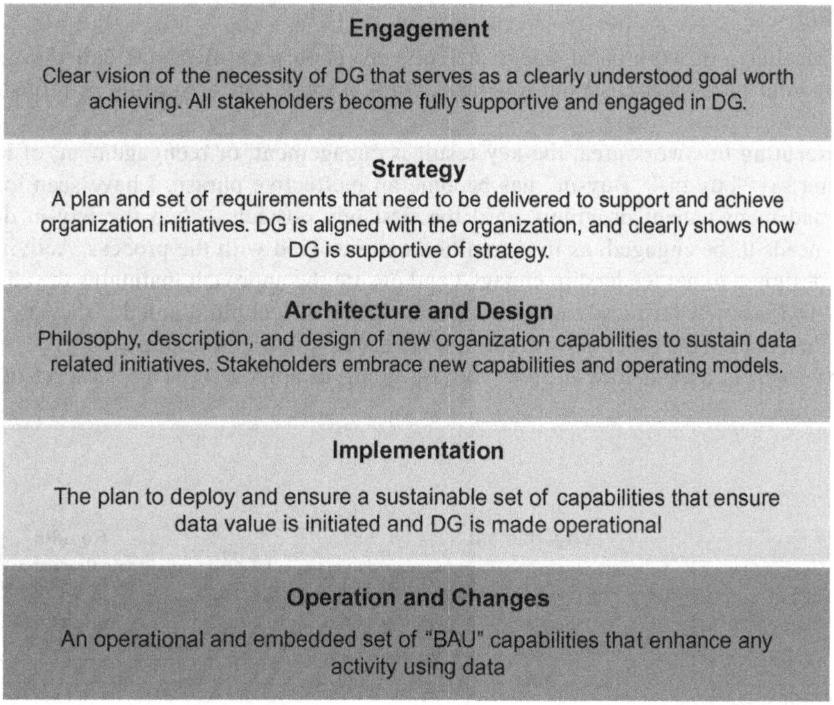

Engagement

Clear vision of the necessity of DG that serves as a clearly understood goal worth achieving. All stakeholders become fully supportive and engaged in DG.

Strategy

A plan and set of requirements that need to be delivered to support and achieve organization initiatives. DG is aligned with the organization, and clearly shows how DG is supportive of strategy.

Architecture and Design

Philosophy, description, and design of new organization capabilities to sustain data related initiatives. Stakeholders embrace new capabilities and operating models.

Implementation

The plan to deploy and ensure a sustainable set of capabilities that ensure data value is initiated and DG is made operational

Operation and Changes

An operational and embedded set of "BAU" capabilities that enhance any activity using data

FIG. 6.3

Data governance activity areas and primary outcomes

Each area can build upon the previous one. However, the steps can also be conducted as a "stand-alone" process if the required artifacts or information for that step are already available; for example from a data-related effort. All of the possible activities are in a reference table in the appendices.

Helpful hint

Think of the framework more as a "checklist" than a "do list." In aviation, a checklist is not a document that tells pilots what they need to do. After all, they are trained do carry out the task of flying an airplane. The checklist is an artifact that is used for two reasons:

(1) Confirmation—did they do what was required in a certain scenario, e.g., pre-flight, take off, climb, cruise, land?

(2) Reminders—in the event of something that is unusual or new, what are some things that have to be done. For example in the event of a fire, certain systems are turned off, passengers are managed certain ways, etc.

The key to using the framework is understanding your scenario and selecting (or building) the right checklist. As each section is reviewed, this chapter will also provide guidance for low profile vs high profile approaches.

Engagement

The first critical area of work is Engagement. This area is purposefully NOT called project initiation. Getting your plan (checklist) put together is only a small part of starting or refreshing a DG program.

When executing this work area, the key result is engagement, or reengagement, of leadership. Note I do not say "buy-in." "Buy-in" has become an ineffective phrase. I have seen lots of buy-in for DG and management programs until the first few obstacles, then the buy-in disappears. Leadership needs to be engaged, as in committed and involved with the process. Activities in this section are designed to get leadership engaged and ensure the approach maintains the engagement. Reminder—DG is a new business capability. Leadership will feel blind-sided if they are not aware of this and then realize the potential weight of DG at some point in the future.

The next result is a clear idea of how you are going to approach the next aspect of your DG (Fig. 6.4).

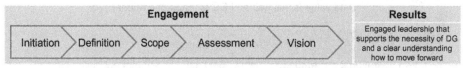

FIG. 6.4

Engagement work area

Considerations

Many readers will be trying DG a second time. This section is used to reengage the old, or new players. If you are starting from scratch, you need to lay out the foundational approach and determine the tone of subsequent iterations.

You also need to determine if you are going to be low profile, noninvasive,[5] or moderately invasive. Or maybe circumstances require a more central, aggressive approach.

Key point

Your approach means considering many aspects of your organization. There is a temptation to make everything noninvasive and try and keep it that way. Remember, at some point DG needs to change enterprise data behavior. If you can do that without ruffling feathers, great. This set of activities will help to look ahead and figure out what approach is practical.

If you do decide on a low profile or noninvasive approach, bear in mind you still need a strategy, a plan, a design, and possibly limited assessments. Please apply some thought to the details. It will pay off.

---

[5]Seiner, Robert, "Non-Invasive Data Governance," Technics Publications, 2014.

Also consider the origin of DG. For example, right or wrong, most DG programs get started within an information technology (IT) area. If a CIO is gung-ho about cleaning up the treatment of data and making it a powerful asset, she had better verify that the scope of DG includes the creation and enforcement of broad-spectrum policies. And have a culture and political environment to support her.

If an organization is highly regulated, then the compliance area needs to be brought into the DG effort. Your degree of central influence will depend heavily on the degree of compliance involvement.

Defining exactly "what" is governed is also of key importance. For example, are any business areas exempt due to regulatory reasons? Is there a division that, due to its business model, DG would not be helpful? For example, a client had a line of business that dealt entirely in research, so experimental data and research papers were the core information assets. Those folks already took very good care of their data!

Besides scope restrictions, you may need to consider factors that require a larger scope than initially considered. What about business market factors? A DG effort attached to a master data project may need to consider a greater scope if a company's market share is suffering and poor-quality data is a contributor. If your company has recently completed or is in the middle of implementing an enterprise-level application project, such as SAP or Oracle ERP, then your DG effort will need to cozy up to those programs.

The intensity of DG is part of the scope decision. Are the information principles that will arise from the DG effort required to have the weight to cover an entire organization? The same decision goes for policies. Your DG program will create new policies and you need to decide to what levels of the organization you will extend those policies. If appointing individuals with new roles of accountability, or decision rights that are new to your organization will be an issue, then your human resource area needs to be considered as part of the project scope.

It is not a trivial matter that the scope of DG is set by the nature of an organization (i.e., the methods used to set and enforce policy and rules, how decisions are made, and who makes them). If an organization has a culture of accountability, then the scope of DG can be broadly stated. If the organization has operated without blatant accountability for information technology and data assets, then DG scope must be stated very specifically, and mention that accountability will be entering the organization's lexicon.

Many other factors influence your approach:

1. Are you doing DG as part of a single MDM effort? If you are doing a typical project to consolidate customer data, you may have to focus your initial governance on the MDM event. Your organization may not yet have an appetite for enterprise DG. You will execute a lot of the framework, albeit on a more limited scope. This does not mean you treat the DG program as a stand-alone effort. Often, there is another MDM project on the heels of the first one (assuming they are successful). Then you will immediately see why DG needs to be treated with an enterprise perspective, regardless of its roots.
2. Do you work in a very large company? If so, my guess is you have simultaneous instances of DG percolating. They may not all be called DG, but they are there. A uniform process allows various efforts to leverage and combine their efforts under a common protocol.
3. Do you have a formal EIM program or information management (IM) area? If so, you will execute this approach probably once to stand up the larger DG "area," and then several times as you support various projects requiring DG.

Lastly, remember you defining DG for an enterprise. That means, start with the whole thing, and only reduce scope for specific reasons. There is no such thing as departmental governance. It is a contradiction in terms.

Activity

The engagement activities are your "get organized" tasks. Make sure you have a program first. Obtaining explicit approval to embark on DG is key.

Anyone who has done any kind of program or project knows you need to start with an understanding of scope. It is no different for deploying DG. After all, there is a great likelihood of affecting several segments of your organization. Identifying what is governed, what areas are involved, and what business capabilities will be exposed to DG is important.

Once you identify and apply constraints you can define a formal scope, and develop the program roll-out plan.

An important step, regardless of your style of approach, is the team that will be getting the program energized. This is not a project team, but a program team. That may affect who is on the team. Certainly, there needs to be resources that can interact with all level of stakeholders and politics.

Once scope is understood—and approved—then the new DG team can move on to the required assessments. Unlike assessments done for data quality or enterprise architecture, the DG assessments are focused on the ability of the organization to govern and to be governed. Use the alliterative phrase "capacity, culture, collaborate." That is extremely important to determine the current state of the mechanisms and processes an organization will be changing as DG rolls out.

"Capacity" refers to the capacity to change. *Desire* to change should never be confused with the *capacity* to change. For example, the IT organization at a past client knew that data quality was the number one obstacle to developing a customer master data management (CMDM) architecture. Business users across the board openly acknowledged that customer data was pretty awful. The project was stalled. Many of the processes to correct the problems were designed, but nothing was happening. The root issue was that no business area wanted to be the first one to assume the new discipline required of the CMDM solution. In fact, it did not take long to determine that not a single department was able to embark upon the required changes without major upheaval. The corporate spirit was willing, but the corporate flesh was weak. It took a major effort to prepare the organization for the required changes.

"Culture" is the number one challenge of DG. However, you cannot say, "Yep, let's manage culture!" and expect to be covered. All organizations have a different way or style of using data and information, even within the same industry. That is, they use data and information differently. Since the ultimate goal of a DG program is better data management (DM) resulting in better information, we certainly need to understand where the organization is *now*. It is good to know an organization's current maturity and how it deals with data in the present. There are numerous maturity scales that can be used to articulate where organizations are in terms of data use and management. Culture is never really changed. Data behavior is changed. Culture is accommodated and leveraged.

"Collaboration" refers to the assessment of an organization to work cross-functionally or to work on a task using teams made up of representatives pulled from various business segments. Granted, this can be considered part of the culture. However, when collaboration enters the DG deployment picture, it is a discipline that requires a thorough understanding of an organization's ability to work collaboratively. A readiness survey can draw out where an organization is in terms of being able to work on data issues.

Based on the three "Cs" described earlier, the assessment phase for DG deployment entails three types of assessments. Whether you do all of them or only a portion depends heavily on the origins of your DG effort. Fig. 6.5 shows what you need to consider along with the three assessment types.

Regardless of what direction, it is perfectly fine to mix and match these assessments. Often the "Change Capacity" is combined with the "Information Maturity" survey, usually due to restraints within the population being surveyed or assessed.

What types of Assessment are needed?

Potential targets of DG:	Assessment types:		
	Information Maturity	Change Capacity	DM / DG Readiness
Support MDM, Data Lake, Analytics, or other large structured information project	Yes, if it has not been done as part of the project		Yes, MDM is, by definition, cross-functional
Support artifical intelligence, machine learning, or data monetization	Yes, if it has not been done as part of the project		Yes, these advanced technologies require organizations, to be "data driven literate"
Support document management, or other unstructured information project	Yes, especially in the context of document management		Yes, document and content management are, by definition, cross-functional
Support data quality, broad version	Optional, the data quality effort is usually focused on creating better data which changes maturity anyway	Yes, if not already done as part of DQ effort	Yes, data quality changes are data behavior changes
Start DG as part of enteprise data or architecture strategy	No, if it has not been done as part of the program already it is not going well		
Support data quality, narrow version	No, really low profile means data maturity is secondary to results	Yes, but only on the initial stakeholders	No, it does not apply
Start DG as a low profile, minimally invasive effort with a Use Case	No, really low profile means data maturity is secondary to results	Yes, but only on the initial stakeholders	No, it does not apply
Start DG as a standalone program	Yes, but why? Stand-alone DG is usually really a form of doing a formal EIM program. Better double check what it is you are trying to accomplish.		

FIG. 6.5

Assessment types

Helpful hint

Collaboration is a word that is becoming as cliché as "culture change" due to overuse. Much akin to "governance" and "culture change," it is a term that is easier to understand than to implement. Remember, the reason you are talking about all three of these terms is that your entire organization is realizing that the way things are being done is *not sustainable*. That means retraining, learning, changing abilities, and adopting new philosophies.

The "Vision" activity is executed to demonstrate to stakeholders and leadership the definition and meaning of DG to the organization. The goal is to achieve an understanding of what the DG program might look like and where the critical touch points for DG might appear. Those new to DG but aware of other strategic program processes may initially say this step is superfluous if the organization is totally on board. However, experience has shown this is a dangerous position to take. It turns out that until you show some sort of "day-in-the-life" presentation, many people do not comprehend what DG means to their position or work environment. In the context of DG, this phase may appear to be more of a conceptual prototype.

Since we are creating a very high level, or notional, representation of what DG could look like, you need to translate scope into a definition of DG that is suited to your organization. Then form that definition into a clear simple representation of scope and impact. You may even want to take a run at a notional roadmap with a comparison of current state to future state. At this stage, you need to do whatever (emphasis on "whatever") it takes to continue to draw more and more stakeholders into accepting the vision.

The vision shows stakeholders and leadership what DG will look like. This means a bit more than a one-page picture, although that is important, too. There may be need for a formal mission statement, and both vision and mission are defined in detail in the coming sections. A vision establishes a picture of where an organization would like to be at a certain point in time in the future. The mission talks about how to get there. The goal is to convey understanding and comprehension of what DG means and what the organization wants to do to get there. This vision reinforces the fact that the business of enterprise information asset management is the business.

"Vision" can be an abused term. It implies fluff and waste to many disillusioned executives. With DG, however, there is a profound need to convey the "big picture." Earlier in the book we mentioned the need for organizational change management. A key aspect of a change program is maintaining a future vision in front of those undergoing the changes. Change does not happen among humans without some view of the big picture. This is your goal for the vision phase. What will a "day in the life" look like when DG is activated? What will be visible? What business goals will be more achievable?

Strategy

The Strategy part of the framework is where the long view activities take place, if needed. Key outcomes are the support of the value proposition, and alignment of DG with strategic requirements of the organization. Frankly, some type of strategy might be useful on even the smallest, initial DG efforts. Even the smallest effort should be able to point to our support for organization goals.

The strategic tasks also encompass defining strategic requirements for DG. This is a topic I have had some debate on with other practitioners. I feel it is important to point out the large strategic data areas that DG will be touching, *if they are known*. For example, if you have decided to start with a low profile, you will need a use case or topic for your initial effort. This activity covers the gathering of requirements so you can find the use case, or initial starting point. If the use case is data quality, then your strategic requirements for data quality can be studied. If you are going to tackle reference data as a starting point, then the drivers of the reference data (analytics, or BI for example) can be studied for data requirements. Also, obvious DG capabilities that will be required for the use case can be brought forth. The main theme at this point is WHAT needs to be there. Anything is fair came if it is obvious—data areas, metrics, capabilities, etc. (Fig. 6.6).

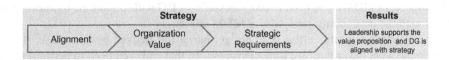

FIG. 6.6

Strategy work area

Considerations

Alignment refers to the direct linkage of the efforts to manage information assets to business strategies and measuring these information and knowledge projects against the anticipated benefits.

Therefore, the DG rollout team needs to make sure that the link between data assets, business strategy, and DG are apparent. This provides the foundation for measuring success, prioritizing capabilities, and the baseline for sustaining the DG effort. Regardless of the DG approach, there needs to be a conscious consideration.

Organization value (of DG) develops the financial value statement and baseline for ongoing measurement of the DG deployment. The DG team will examine (in more detail) the business strategy and goals and develop a link between DG and improving the organization in a financially recognizable way.

Two aspects to this area merit careful consideration. First, you need to consider what else is going on in terms of managing information as an asset. If there is an overall DM program or organization, or there are data intensive programs like MDM, analytics, or data quality, then some of the effort described in this phase may have already been done.

Helpful hint

When you are around the vision or business case activities, you will undoubtedly encounter the first layer of resistance to DG. You will attempt to present to an executive level and three things may happen:
1. A lower level will be told to deal with it. The executives will be too busy.
2. Your sponsors or business representatives will get cold feet when it is time to educate in an upward direction and will dilute the message.

3. The executive level will humor you and sit through a presentation, ask some good questions, and then forget you ever met.

All three represent a lack of understanding. Experience has shown that the highest levels of resistance are usually put forth by the organizations most in need of business alignment! However, repeated education and reinforcement of the message accompanied by some good metrics will start to open doors. You may have to revisit and repeat vision and business case activities over a period of years as you penetrate more areas of your company.

Remember this is a mandatory step in deploying DG at some point. It is good news if some or all of it was performed as part of another effort. Even if there is an associated program (like data quality, advanced analytics, or MDM), you need to take stock of how DG will support the business, even if it is indirectly through the data quality, analytics, or MDM efforts.

Maybe the initial, low profile effort does not get into a full-fledged business case. But you still need to determine what the criteria are for DG success. To that end, you need to perform this activity to the extent required to provide the baseline for determining DG performance metrics and measures of sustainability.

However, larger efforts or more visible applications of DG (artificial intelligence [AI], advanced analytics, MDM, etc.) require a business case. This is a great opening to delve into the accumulated data debt of an organization and start to inform of the long-term consequences of continuing data silos, neglected data quality, or misalignment of IT and data projects.

The alignment and value activity lead straight into identifying strategic requirements. These are preliminary, broad things that you know will need to be incorporated, regardless of your approach. For example, if you need to be low profile efforts, you can decompose your alignment and value results into determining a use case that will meet the requirements of low invasiveness or visibility, yet still prove value. And, your efforts will be accretive to an enterprise DG program. (No throwaway proof of concepts permitted!)

The requirements tasks will also identify obvious, required capabilities. Capabilities are new in this edition. "A business capability is *what* a company needs to do to execute its business strategy."[6] It represents the ability of an organization to perform some activity or process that results in an outcome of value. Capabilities are best presented in the context of WHAT happens and in terms of business outcomes and value. Capabilities are used because it gives the aspiring DG area a very common language to communicate with business areas. In addition, DG is, in itself, a required business capability.

The final type of requirement to be aware of at this point are your data, or information, principles. The DG team identifies, documents, and vets the core organization principles that will need be adopted. Depending on scope they are for a use case, or the enterprise to manage information as an asset. Without a discussion of principles at this point, the DG effort gives away a key element for success. For a smaller effort, or noninvasive effort, there still needs to be discussion as to what principle applies to the use case. The groups standing up may not be able to deploy an enterprise principle, but they must be aware of the philosophy they want the organization to adopt. This influences road map, training, and change management activities.

---

[6]Wikipedia.

Helpful hint

One bit of feedback from edition one was not enough book area devoted to selling DG. As you have read, I added more in that area. However, you should not have to sell a required business capability. Selling something that is required to make your strategy happen is easy, in fact, selling DG is not even the right way to phrase it. You need to focus on education—not selling. If you have to sell the idea, then your leadership does not understand why data itself is important yet. That is not selling. That's education.

Activity

The alignment activity is a deliberate connection of business goals and initiatives to DG and DM. If you are doing DG to support another effort, like AI, MDM, etc. there may be data to tie DG vision to business needs. Additional details about business goals and objectives are turned into specific value statements where DG enables positive change. For example, the number one area where DG can assist most companies in the BI and reporting areas is to ensure business alignment with BI initiatives and technology. So there needs to be a call out of clear business objectives associated with the BI efforts. If there is no other source of an IM business case, then the DG team needs to execute this activity.

I have often witnessed a scenario where an IT department starts an information-centered project. Most notable examples are stand-alone MDM efforts where the CIO tried to integrate core data as a technology effort, or a Big Data, data lake structure is put into place without any regard to data quality or lineage. The DG team needs to fully understand business needs and isolate those actions where correct and well-governed information will help the organization achieve its desired results. This may not be a trivial effort where organizations need to do a lot of things fast with data or are undergoing multiple large projects. It will mean doing an exercise to map strategies to information projects, an activity that is often met with interrogation as to "why" or outright resistance.

The value of DG is determined when the DG team identifies specific financial numbers and determines what business metrics will indicate the success of DG. This is also a good place to show the cost of nongovernance or continuing to use information in a poorly managed fashion. This task means getting into true business benefits—reviewing the returns for cost savings, data monetization, or new products and customers.

Any activity on business case or alignment leads into an initial scan of requirements that will shape DG. The definition of key data areas, metrics, or data issues will allow you to organize the first cut at what obvious capabilities need to be fielded. Do not start with specific data sources—start with what business goals DG will help achieve. Then move into specific business events, requests, metrics, statistical models, and regulatory areas. If you are deploying DG as part of Analytics, AI, MDM or similar program, these elements should already be available. If they are not, then this is the opportunity to orient these efforts and pull them back from being technology-only efforts. For example, if an MDM program is talking only data sources and file clean up as requirements, or an advanced analytics area is only talking about trial models with a "we know it when we see it" approach, these efforts are probably derailed.

Architecture and design

This work area contains the largest and most impactful set of activities of the framework. I can confidently say that every single DG effort, new or renewed, narrow or broad focus, will need to dip into this set of activities. The tone in this set of activities is to get stakeholders fully engaged, determine what the moving parts are going to be, and then design the moving parts. This section will touch people, process, technology, and data (Fig. 6.7).

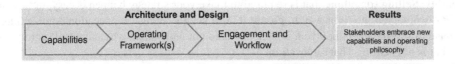

FIG. 6.7

Architecture and design work area

Considerations

Why does every type of DG effort go through this activity area? Simply because this is where we design the solution, regardless of how small or large, how invasive or not. Once we know what DG needs to accomplish (alignment and requirements) then we need to identify WHAT and HOW. Obviously, this is not a new means of arriving at a solution. For DG, however, many program teams are in new territory, having worked with discrete programs or computer systems.

Capabilities are used because they are the language of business and enterprise architects. Since DG is usually a new business capability, it is important to keep these areas engaged. If these areas do not exist, then the DG team needs to assume the temporary role of enterprise architect. Even if a DG program is in place and you are reenergizing your program, shifting toward capability thinking will strengthen your engagement with stakeholders.

You are building a program that requires a framework for operation. That means some basic blocking and tackling in terms of Management 101. But this is *not* an exercise in scraping together a few organization charts. The term "operating framework" is purposely used instead of "organization chart." Given that the goal is to eventually blend in with ordinary day-to-day behavior, you will rarely develop a large separate DG organization. There will always be a small virtual function of DG visible, but does any organization need a stand-alone, permanently funded DG "department"?

There are detailed processes that are designed that make the WHAT of a capability become a HOW. Also, this blends in workflow with how DG engages with its constituents. Obvious areas, even for smaller efforts, are affected systems and data users. For more strategic DG programs, engaging with strategic planning, budget planning, and compliance areas will require some process definition. Avoid the common mistake of blending DM processes with DG processes. This is usually because the initial staffing of DG is drawn from information areas. Often the initial DG staff are told to "fit it in" to their current roles and responsibilities.[7] It is challenging for these individuals to maintain the separation of duties while creating embedded organizational processes, new roles, and a sustainable program. Physical separate areas of DG and DM are often not possible, but please maintain the spirit of separation of duties (the "V"). See Fig. 6.8.

---

[7]Kudos to those IM staff I have worked with over the years who have all had to do double duty. There are many hard-working people in data/information management, and I have never seen leadership allow the designated DG deployment team to offload their current duties. Of course, it drags things out, but they hang in there. As for management demanding the double duty without additional incentive, while at the same time saying how important DG is, well...

Lastly, pointing out new roles and who needs to do them is considered at this point. For the low-profile approach, this may only be one or two individuals. A broad-scale, compliance-driven DG program, or large MDM effort may require a larger effort to identify responsibility and accountability. The same applies to high impact, but localized efforts like advanced analytics. Whoever handles the data and verifies its readiness for analytical models is critical, and regardless of how many data scientists are involved there will need to be some formal role and responsibility definition.

This step also entails identifying the stewardship/ownership/custodian population. Please note the mention of stewards and custodians and all similar roles has been delayed until the functional design is completed. It is not effective to mention these roles earlier in the process. It places people in a position of feeling they need to do something, but that something is usually ill-defined until this point. It also avoids spewing the whole stewardship vocabulary around before you have actually defined what that means for your organization. Be patient, designate roles and responsibilities, and only then assign the appropriate label to a specific catalog of duties.

Activity

The capability activity needs to solidify the initial capabilities and add any required detail. These need to be confirmed and aligned with the corresponding business needs. Should there be any talk about technology to assist in DG, the DG team will have sufficient knowledge to start to specify types of technologies.

The operating framework starts with getting the required processes identified. All organizations do "stuff." This is where (usually using a list of generic processes) the DG deployment team determines the core list of what DG will be accomplishing. In essence, you add the details to evolve the V—often by developing process models (think flow chart, swim lane presentations, etc.). The team also points out where current business processes are changed. For example, we have often found that a detailed presentation of the DG issue-resolution process is required (i.e., how an issue is identified, recorded, promoted, and resolved—once a client even designed a "911" process for emergency attention to data policy transgressions). Lastly, do not fail to consider the IT areas in addition to changes in business-user activity. Processes and methods for developing and managing computer applications will also change.

In a similar fashion, the DG team gathers (or helps in defining) DM functions. Remember not to blend the two areas; it will result in confusion and a loss of effectiveness of both areas.

Processes that are direct manifestations of principles need to be tagged as policies. DG is an oversight capability, therefore there will be policy to implement.

The essential lists of DG and DM processes are not at all useful until the DG team identifies who does what, and what the various levels of responsibility are. The DG team examines the functions to identify where responsibility and accountability might need to exist to ensure sustainability of DG. This is more of a first pass so management can understand the change potential and be able to consider the new DG processes and framework in context and in an intelligent manner.

All of this activity is pointed toward meeting a use case for DG or building out a larger scope program. In either case, there may be an initial version, and a later version, of the operating framework. Rarely does an organization start on Day One with its ideal state model in place. Therefore, it is necessary to consider if there is a minimum viable state in the interim—a point at which DG could operate, keep the organization engaged, add value, but still have some growing to do.

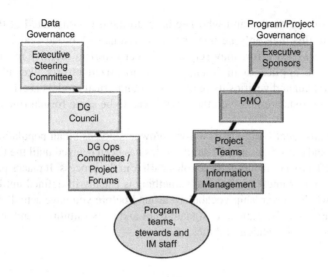

FIG. 6.8

Separation of duties example

By this point, it is very important to educate and present the new operating framework, responsibilities and accountabilities to management. Do not be surprised if there is some back and forth at this point as reality settles in to middle management (i.e., someone is going to be held accountable for data). There will be feedback on any principles (reflected in your policies and processes).

Implementation

This is the step where DG plans the details for the "go live" events of DG. The team will define the events that take the organization from a nongoverned to a governed state for its data assets. In addition, the requirements and groundwork are laid to sustain the DG program (i.e., detailed preparations to address the changes required by the DG program) (Fig. 6.9).

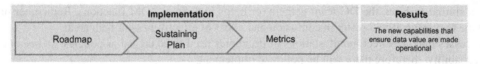

FIG. 6.9

Implementation work area

Considerations

For low profile efforts the DG team is planning to roll out a limited set of capabilities and deliver a use case, often in conjunction with a DM solution. This means the roadmap is most likely a project plan. However, any other approach means the team may be creating a tactical plan that will take several years. This is not a light-duty activity. The "roadmap" that is produced from this step integrates DG activity with other projects and initiatives. In fact, most of the time you will need to "piggy back" DG on other efforts (unless you are doing DG because of another effort). There may need to be a tactical aspect and a longer-term aspect, presented simultaneously, depending on scope. The key activity at this time is figuring out what the incremental approach will be, and the size and make-up of the increments.

Some sort of formal plan for sustainability is required, regardless of approach. It may be a 30-min session to identify training and some hand holding, but do not skip planning for changes. Conversely, I have seen the need for a plan that required a five-person formal organization change team for an organization that had to govern and manage its data or be dissolved by regulators.

Regardless of how the change management and rollout activity is developed, make sure there are frequent checkpoints and opportunity for feedback. Again, you will be changing behavior. Don't let circumstances and lack of attention create a situation where the organization can make an excuse to "defer" DG.

Make sure that output (communications and training plans as well as a roadmap) are all tuned to the organization's culture. Too often inexperienced teams deliver generic results from this step (i.e., a few newsletters, a mass "training class," and a one-page Gantt chart). Frankly, most of these will be overlooked. The DG team will receive a full-on response of "been there, done that." These tasks require some creativity.

Lastly, consider what needs to be measured going forward. A low-profile effort may only need an indicator of effectiveness and some feedback to management of reaching project goals. Broader approaches may require a host of metrics. Lastly, always hold up the concept of data debt when looking at metrics.

Activity

DG needs to be woven into the fabric of everyday business. Therefore, the DG team needs to review, align, and if possible, jump on board with other efforts. Remember, if there is a program "sponsoring" DG, like MDM or advanced analytics, you need to integrate with that program's plans. For a low-profile approach, make sure there is sufficient detail to easily manage and coordinate DG activities with any other interfacing project. Broader approaches will require definition of rollout increments and tactical and long-term views.

The team needs to define the sustaining requirements, that is, what will be required to keep DG sustainable. There may be many cultural elements that need to be addressed if there is to be a successful DG rollout. This activity determines these elements and how they are coordinated. The team reviews the change capacity assessment, stakeholder analysis, and any other findings gathered during the previous activity with the intent of developing the requirements for ensuring the DG program is sustainable. Try to avoid the deployment of, well, almost anything, without considering what will need to happen 1 or 2 years down the road. In addition to training and communications activity, which are obvious, there will need to be an ongoing measurement of the attitude and morale of the DG team and stakeholders. Change efforts require long-term sponsorship, so the DG team will be looking for an individual to act as a change sponsor.

The requirements for change lead into the development of a formal change management plan. This will entail metrics to measure change (not to be confused with the metrics for DG effectiveness) and the development of reward structures and compliance activity for stakeholders who are moving into a world of well-managed data assets. For larger scoped efforts, the change management plan is fairly detailed and should encompass a period of 1–3 years.

Once the requirements for change are understood, the details of the rollout of DG are put together. This means blending the sustain plan, and the details within the roadmap. The actual steps to start DG, including details for the stewards and custodians, are presented.

Lastly, you need to measure what you manage. DG requires some sort of metric-based feedback to ensure its continuity. If you cannot show demonstrable effect, it's too easy for naysayers to slap the program down when the changes start to take effect. Therefore, the DG team needs to define some solid progress metrics and reports.

Helpful hint

The successful deployment of DG will be viewed as yesterday's news unless it is kept visible (and someone important gets credit for its success), and that is the purpose of the sustaining activity. We approach the planning and rollout of DG with the viewpoint that modern organizations, especially modern corporations, have the attention span of a 2-year-old. This may or may not be true, but it helps with the planning.

Operation and changes

This work area represents operation of the DG capability and execution of the activities related to sustainability. In essence, once you have started to sustain DG, it never stops. Until DG is totally internalized, which may take years, there will be the need to manage the transformation from nongoverned data assets to governed data assets. There is no stop date (Fig. 6.10).

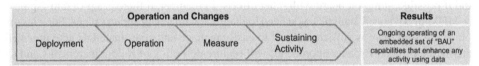

FIG. 6.10

Operation and changes work area

Considerations

The DG team (actually the entire DG framework) starts doing DG. That means the operating framework is, well, operating. There can be a few activities, such as modest capabilities with an initial use case, or a broad implementation of several capabilities. Any desired technology is procured and put into operation. Training and establishing work groups is usually underway at this time. Regardless of approach, any successes need to be widely broadcast. Any challenges need to be dealt with rapidly.

Any mechanisms for metrics get deployed as well.

Of course, for broader efforts with a roadmap, the team follows the roadmap and diligently reports progress against the roadmap.

The team works to ensure the DG program remains effective and meets or exceeds expectations. At times, there will be reactive responses to open resistance. There will be proactive tactics to head off resistance. The main emphasis will be to ensure that there is ongoing visible support for DG.

Any material on sustainability provided in this book is really material based solely on the business discipline of culture change management. In our practice, we have evolved to using the term "sustain" simply because it's more understandable and accepted than "culture change."

DG is not self-sustaining. First and foremost, this must be accepted. While the net cost of DG, over time, is zero, there must be the understanding that formal activity is required to ensure you reach the zero-sum state. Remember that the eventual goal is to make DG institutionalized and not a separate concept. This phase should also reflect periodic replanning, as personnel and business needs will change. The DG program needs to adapt without losing focus.

Activity

At last. The DG team, along with whatever the appropriate project teams and DG forums are, actually start to "do governance." Whatever initial groups have been designated (via the roadmap) are indoctrinated into new processes. This means, of course, training and communications. It also means publishing many of the artifacts that have been developed (e.g., guidelines, principles, policies, etc.). The DG stewards and owners who are responsible for reviews and audits start these activities as well.

The team starts to execute whatever type of change plan has been developed. All of the activity defined to address sustaining DG occurs here: communications, training, check points, data collection, etc. Any specific tasks to deal with resistance can be placed here. Over time, training and educational material will require updating. Additional staff will require orientation. Management will need to hear about the bright spots and not-so-bright spots. All of these elements of the culture change can be listed in this activity. Initially, the most effective tasks to be defined are the ones where resources need to be involved in communicating, training, or addressing resistance.

The DG framework needs to be scrutinized for effectiveness. A separate forum or a central DG group will carry this out if one exists. Principles, policies, and incentives need to be reviewed for effectiveness. Even leadership and sponsors need to assess if the effort is large enough.

Low profile efforts just need to verify that any new capabilities are being used. A good example is the initial use and population of a glossary. Far too often only a few individuals stay with glossary use.

Be sure to separate effectiveness of the framework (the federation of responsibilities and accountabilities) from the general effectiveness of DG. This will entail data collection and the generation of metrics that report on effectiveness of policies and standards, as well as the activity of designated stewards and owners. Focus groups, interviews, and surveys are common techniques used to assess how the rest of the organization views DG. If changes are required in DG policies, then this activity triggers the necessary adjustments.

Summary

This chapter provided an overview of a series of activities that are useful for "standing up" DG. The following chapters will provide specifics on the tasks and work products necessary to deploy DG. The main concept to take away from this chapter is that DG deployment, while being programmatic in nature, still requires a defined process and rigorous management. There is no one way to "do data governance." It varies by scope and within organizations.

Core success factors

There are three core success factors to reinforce at this point:

1. DG requires culture change management. By definition, you are moving from an undesirable state to a desired state. That means changes are in order.
2. There is no DG "organization." It is a business capability, and is not tied to departmental or organization designs. Ideally, in most organizations DG will end up being a cross functional, virtual activity.
3. DG, even if started in a low profile, even stealthy nature, needs to be tied to an initiative or support a business capability.

Much of what has been presented is not rocket science. It never hurts to revisit basic "blocking and tackling" activities when you might be new to standing up new business capability.

Essential questions

1. DG needs to follow a precise waterfall approach. True or false?
2. DG does not adapt to Agile methods. True or false?
3. What are the basic considerations when determining how to tackle the deployment of DG?
4. Explain what is meant by the statement "leverage your culture, do not change it?"

Engagement

Chapter Outline

In preparing for battle, I have always found that plans are useless, but planning is indispensable.
Dwight D. Eisenhower

Data Governance. https://doi.org/10.1016/B978-0-12-815831-9.00007-2

Overview

Starting the deployment of your data governance (DG) program entails much more than standard program startup activities. Simply stated, it is difficult for most organizations to get started.

I'm taking a slightly different tone from the first edition. For that, I started with the assumption that DG was good to go, and we plunged into "doing stuff." Experience has provided additional considerations. It is critical that organizations not only buy in—but actually engage with a new way of handling data. "How do I sell the need for data governance?" is a frequent question. The answer has changed over time from "build a good business case" to "build the business case and make darn sure your sponsors are engaged from the beginning." Why? Because "buy in" is easy. It means muttering a few words that you are supportive of the cause. *Engagement* is different. Many organizations can tackle large projects. But initiating a new program is a different animal.[1]

Traditional activities such as timelines, participants, project administration, and communications still need to happen. But the key output for this area of work is a solid ENGAGEMENT of all stakeholders. We will cover low profile and high visibility efforts. But engagement is the main goal regardless of your approach.

This chapter, and the ones to follow, provide details on significant themes. Depending on your required approach, you will need to use some or all of the activities in each area. The activities within the areas are conceptually connected. The need to be agile, or iterative or less invasive requires that we step away from a pure methodological mind set. Your roadmap from your initial iterations will mix and match material to provide the linear project plans. I will show examples.

There will also be the need to execute new activities that are unique to DG. This chapter and the others will cover those in detail. I will also introduce the second case study, which will reflect a DG effort requiring a broader scope and approach than Rocky Health Care.

Please do not assume the initial tasks are a casual exercise. The typical program/project plan deliverable from this phase has averaged 400 tasks. My practice has produced DG deployment plans that range from a low-profile plan of about 100 tasks that last a few months, to plans that span 3 years and contain over 1000 discrete tasks.

This activity area is not supposed to take a long time. Many of the tasks can be done in a few hours or a single meeting. I may be long-winded in discussing them but remember these are all checklist items. Many can be combined.

It is critical to comprehend and convey the type and amount of activity that can possibly take place, and how the workload will be addressed. The vision of what is possible needs to be clear enough to keep all stakeholders locked onto the effort. The assessments, if done, need to quantitatively show the capacity of the organization to deal with the new, or reenergized, program.

Hence, the quote at the beginning of the chapter—the planning activity sets the tone and the team. Perhaps the most well-planned activity in history was the Operation Overlord invasion of Europe in WW2 (sometimes referred to as D-Day). That event took 2 years to plan. The invasion was successful, but the plan was quite fluid once the event started.[2] Therefore, your plan itself will change over time, but the focus and artifacts will help sustain the DG effort.

---

[1]Don't be tempted to use the familiar public speaking joke about commitment to convey the difference between buy-in and engagement. Basically, it goes that in the context of breakfast, chickens are involved, but pigs are committed. Of course, you need to understand a western breakfast of bacon and eggs to get the joke. And then you risk the parties that you want to be engaged feeling that they have been called a pig. In addition, this joke ignores certain religious beliefs. So, find another story. Somebody told all this to me. I would never have said this. Just sayin'.

[2]Near as I can tell, the planning and execution of the Normandy invasion also gave rise to this quote, "The reason you have a plan is so some SOB can change it." The originator is lost to history, but it certainly captures the essence of DG planning.

Fig. 7.1 shows the types of activity in this area.

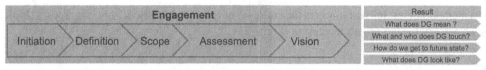

FIG. 7.1

Engagement work area

When you have completed the activity around the Engagement theme, you will have established the meaning of DG to your organization. If you are trying to restart your DG program, don't overlook this. It may be a reason you have had issues. You will have identified all participants, with roles, responsibilities and work assigned. Even efforts with noninvasive or agile approaches must get stakeholders engaged.

You will be able to present a crystal-clear vision of scope and span that are adequately understood, and then produce a plan that will sufficiently guide the team through DG deployment. There had better be a project plan, regardless of your approach. There is a strong case to be made that a low-profile effort requires much more detail on the project plans. Frankly, I make sure low-profile efforts have almost excessive details. This makes sense, as things are happening too fast to ponder what to do next.

Most importantly there will be adequate information available to leadership to keep them engaged. They may want a business case, which can be provided. But there will be adequate vision and understanding of the approach to keep anyone who authorized the program interested.

A final reminder that the work areas are not a recipe, where Engagement tasks are all done before the next area (Strategy) starts. The various activities within Engagement—Initiation, Definition, etc. are not necessarily in sequence. These are all things that you can do, depending on the circumstances. For example, Strategy and Engagement are often combined into a Phase 1 for smaller or lower profile efforts. Again, think more like a checklist vs a recipe. So remember, this chapter presents a batch of conceptually related activities and the guidance on how to use them (Fig. 7.2).

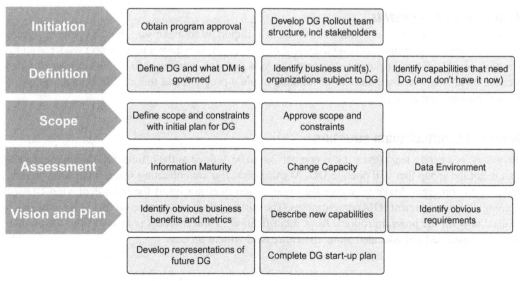

FIG. 7.2

Engagement work area activity

Initiation

In most methodologies, Initiation covers the entire set of tasks to get started. For our purposes, Initiation is just an activity to start program creation (Fig. 7.3).

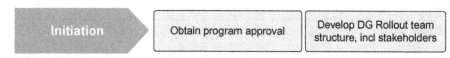

FIG. 7.3

Initiation activity

Obtain program approval

An important, early activity, regardless of size or approach, is to get leadership to say that you can start building a program, not a project. Even if you are going to be low profile, or noninvasive, you should deliberately develop an awareness that if this all goes well, leadership will support a program that will create and sustain DG capability. Later on, the metrics and measure of success will be defined to prove the value of DG, but you need that programmatic awareness now.

Approach considerations

I recommend that every iteration of DG verify support of a program mind-set, even if you are trying a second or third time. If you believe that politics and culture require a lower profile start, still make sure that leadership provides some form of approval. Even if, heaven forbid, someone calls this a proof-of-concept, at least get an acknowledgement that good results will mean recognition of a program vs completion of a project.

Ramifications and benefits

Since so many DG efforts start with one person, it is important to make sure that this person is recognized and supported. Without acknowledgment of a program, you have a new project leader. And that is not sustainable. Formal acknowledgement also gets you one step closer to data governance literacy, because there is now a small amount of thinking about a program and that will lead to acceptance of the new business capability.

Develop DG rollout team structure

Obviously, you need a team (even if it is one person) to be defined as the official DG deployment entity. This is not the group that will operate DG. At this point, you are collecting some smart folks together to define the program. Make very sure the organization does not insert the team deploying DG with ongoing data management (DM) practitioners. The steering body or sponsors at this point may also be focused only on the program rollout. Also, identification of stakeholders is key. Ideally, stakeholders need to be assessed and educated along the entire deployment process.

Approach considerations

If you are on a second or third attempt at DG, there may be carryover of individuals, but there should be a sense that they are willing to go a new direction. The most common root cause for a DG program starting over is failure of leadership to engage fully with the prior attempt, but the significant drivers for DG still remain. The most common reasons are:

1. Failure to fully educate leadership
2. Lack of business alignment (more on this later)
3. Appointing a bunch of stewards day one, then having them do nothing for several months
4. Buying technology, that does not work (because you haven't designed how to use it) then getting heat from management to justify the acquisition
5. Building a DG department vs a capability

Low profile approaches will require a team that can hold its own. Since there is usually only one to three individuals in these situations, they had better be used to the ebb and flow of politics and changing priorities.

A large scope DG program, by definition, means more resources. In this case the DG team lead needs to be strong, and hopefully experienced in new programs (not projects).

The most important team member at this point is the sponsor. A sponsor needs to go past buy in, the need to be fully committed and be passionate about the program. They need to be able to spend some political capital. It is not uncommon to start with one sponsor, and then switch to another, so always leave that as an option.

Ramification and benefits

Large teams mean more diversification of skills. This includes skills such as:

- Write policy
- Design functional models
- Training skills
- Facilitation
- Managing steering bodies

For larger teams you may want to consider a SWOT (strength, weakness, opportunity, threat) analysis in this activity. A SWOT analysis is a well-known technique to assess a team's or an organization's potential. In the case of a team, every individual is assessed by what strengths they bring, weaknesses they may have, and what opportunities or threats participants may pose.

You need to watch for early signs that the team is not taken seriously. Even at this early point, larger efforts will see politics, even resistance at this point. Some other warning signs are:

1. Viewing the DG team as the programmer "graveyard"—usually IT staff are the first members of the DG team. Often, individuals will be submitted for membership because they do not fit anywhere else. This means someone is not taking DG seriously. The DG deployment team needs to be experienced in internal politics, know the players, and be able to think outside traditional information-management functionality. Typical roles are shown in Fig. 6.8.
2. Getting a steering committee that immediately delegates attendance to non-decision makers. Again, DG is not being taken seriously. More education is required, so add it to the plan.

There are two alternatives when the rollout team runs into obstacles. Note we said *when*, not *if*. Here are some scenarios:

1. The Chief Data Scientist raises an issue, and sets off to fix it while the DG program is getting started. Normally, it is something along the lines of "No, you cannot use that resource anymore." But it could be a data issue as well. The DG team needs to submit an issue to the Project Management Office (PMO) or similar body that oversees the data science area. That might be the Chief Analytics officer or the equivalent. This is one of those weird transitional things that happen, but it will certainly occur.
2. If the DG team is part of a larger effort, then they should proceed with whatever staff they are offered, and then build in additional time for training and team building. They either will get the extra time or will have a solid case for getting other people assigned.

Remember that this rollout team is not permanent. They will be able to go back to their prior duties once the program is operational. Also, remember, DG is not an incremental increase in head count. People will roll on and off of DG roles over time.

The key advantage of this approach (a deployment team that transitions to a new group for operations) is the better skills leverage.

Tips for success

If the team is not getting adequate resources, then the DG effort is poorly formed. I have worked with a client who assigned one person to stand up DG in an organization with over 500,000 employees on three continents. The DG program was designed and laid out in a vacuum, so when it was eventually presented for implementation, not much happened.

The deployment team should include some experience in DM and data quality. They will have the skills to recognize data issues. Business subject matter experts (SMEs) and someone with a good knowledge of the applications portfolio are also valuable.

You do not need hordes of people to stand up DG. Even the large organizations only need four to six FTEs to get it right eventually. The key is having a powerful steering body and sponsor.

Definition

This set of activities frames the DG effort in the context of DG to the organization. This means a good definition, and an understanding of business areas and capabilities that may be affected. This is often much more relevant to attaining deep engagement than a financial business case. The final declaration of scope may adjust the business areas and capabilities within scope, but at this point you need to consider everything within the realm of the DG effort. If you are reenergizing a program, separate the new areas being addressed from areas that have been part of DG in a prior iteration (Fig. 7.4).

| Definition | Define DG and what DM is governed | Identify business unit(s). organizations subject to DG | Identify capabilities that need DG (and don't have it now) |

FIG. 7.4

Definition activity

Define DG for your organization

During this activity, the team will work with necessary stakeholders to draft a clear, brief definition of DG, and what data is to be managed. An "elevator speech" is likely to be the most visible result from this step. Experience has proven that a straightforward elevator speech adds tremendously to long-term comprehension. (Yes, there is some marketing going on here.)

Approach considerations

Keep in mind that you are looking for comprehension. It is not going to come easily. Regardless of how wonderful you believe DG is, in concept, remember that nearly everyone you are dealing with thinks:

- It is already being done (they will be surprised).
- DG does not merit the same status as financial controls or compliance because it is just data.

Be simple in your definition; keep it concise and sincere.

You also should fold in what DG will do for the company, using terms such as "ensure value, increase revenue," etc.

Whatever you do, *never* use words such as "better data," "improved decisions," or similar terms. They are vague and irrelevant to most executives.

Ramification and benefits

Starting with a clear, first cut at the definition of DG will result in a smoother process to define a vision. The resulting elevator speech is something that the entire team and, eventually, the DG managing framework should memorize.

If you are starting with a low profile, you may not have to present the definition to very many stakeholders. But you should still have some sort of initial definition ready for presentation.

The business will begin to recognize that DG is *part* of the program to manage enterprise data and information—*not the end.* It is the *means* to achieve the end.

Sample output

Below is a batch of sample definitions, all of which are the real deal and are in use. Note that all of these talk about control, or oversight. They do not mention "doing" analytics, or master data management (MDM). Avoid being thrown in with DM. There will be more on this in upcoming chapters.

- DG is a framework of accountabilities and processes for making decisions and monitoring the execution of DM. *(financial organization)*
- DG is using a horizontal perspective of the organization and focusing on the major "pain points" for our business areas. *(financial services)*
- DG is the orchestration of people, process, and technology to enable the leveraging of data as an enterprise asset. It affects all organizational areas by lines of business, functional areas, and geographies. *(software company)*
- DG is a system of decision rights and accountabilities for information-related processes, executed according to agreed-upon models which describe who can take what actions with what information, and when, under what circumstances, using what methods. *(consultant)*
- To be clear, it is the exercise of executive authority over business data. *(chemical company)*

- DG represents the program used by ACME to manage the organizational bodies, policies, principles, and quality that will ensure access to accurate and risk-free data and information. DG will establish standards, accountabilities, and responsibilities and ensure that data and information usage achieve maximum value to ACME while managing the cost and quality of information handling. DG will enforce the consistent, integrated, disciplined use of information at ACME. *(energy company)*
- DG is the organization and implementation of policies, procedures, structure, roles, and responsibilities that outline and enforce rules of engagement, decision rights, and accountabilities for the effective management of information. *(generic definition used by author)*

Examples of an elevator speech are just as varied:

- DG will support our information asset management, ensuring we maintain our market share and achieve cost targets.
- We are going to support cost management and market growth through more disciplined management and monetization of data assets.

Identify business units (subject to DG)

This area of activity supports scoping, by defining and identifying business units and/or organizations that can come under the oversight of the DG program. This starts the DG team toward understanding the possible span and depth of the DG program. It is part of the definition because it needs to reflect the enterprise. When you actually declare scope, you may need to adjust for political reasons.

Approach considerations

If you are intending to start DG with a low-profile approach, identify the candidate business areas anyway. Remember the typical low-profile effort addresses an area that is favorable to DM, and helps that area adopt a more formal approach to data oversight. This is the heart of noninvasive DG (Seiner). A formal examination of business areas (and capabilities—the next activity below) will solidify the guard rails around the low-profile effort. This may change but will suffice for scoping discussions. Obviously, some explanation of DG is in order before this happens.

This is why the prior task focused on definition. Over the years too many efforts had a scope statement, but still no understanding of what was exactly being scoped. If the answer to any of these questions is "No," then the subsequent activities need to add effort to reinforce DG concepts.

They need to verify:

- Is there a working definition or perception of what DG actually is?
- Is DG truly sold? Is more work required to engage leadership?
- Is there at least a notional understanding of the long-term success factors and impact of DG?

Once you have confirmed the definition, it will be easy to apply it to whatever business units or projects that are in scope.

Ramification and benefits

Once the typical DG deployment team wades into this activity, they are usually entering new territory. They are establishing new concepts and capabilities for an enterprise. The benefit is a higher level of skills. However, I often see DG programs stumble at this point when a leader says "Hey, that is our job." In a sense that is true, but someone needs to lay out the details. So you may need some quick communications to leadership at this point.

Identify business capabilities that need data governance (and don't have it)

This activity is a good time to start planning for obvious areas or functions in your organization that require better data. Focus on the capabilities, i.e. the WHAT, vs where, who, and how.

Approach considerations

Use an existing capability diagram (if there is one) or a standard model for your industry. New or affected capabilities are a serious consideration when you start to firm up scope.

Ramification and benefits

You will most likely still need to identify at least one or two business capabilities, even if a low-profile effort. Any use case will have a narrow scope, but the best way to convey any size of scope is to also convey the capabilities you are going to improve. If you are starting governance efforts like MDM or analytics, it is really critical that you disclose affected business capabilities sooner than later. Most likely those projects have already assumed that some processes will work in a certain way, and DG may affect that assumption.

Enterprise architects usually use capabilities modeling techniques, and they are a good source of material during this and any other capability-related activity.

Scope

The scope of DG is a function of span across the organization and the anticipated degree of penetration into business and information technology (IT) activity. For example, a large financial services organization may require a wide span of DG given the nature of its products and regulatory environment. The same organization may require a very deep level of penetration where DG policies will manifest themselves in all aspects of the business.

All efforts to start, or reinvigorate DG requires a formal statement of scope. For the low-profile effort, the scope will be constrained to a use case or functional area. The context will still need to imply an enterprise capability. But this is where the need to start small becomes clear and needs to be broadcast. This set of activities will tighten up, then confirm those segments of the business that will most likely come under the influence of the DG program (Fig. 7.5).

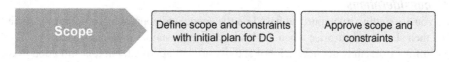

FIG. 7.5

Scope activity

Define scope and constraints with the initial plan for DG

This set of tasks develops an initial scope and plan for the definition and roll out of the DG program. Any required constraints are identified and applied.

Approach considerations

Obviously, the scope determination will be dependent on whether the DG program is based on a program like MDM or analytics, is part of a larger program from a compliance driver, or is a low-profile, noninvasive effort.

Scope can be adjusted here. That is why some planning of the amount of effort takes place. This prevents the DG effort from possibly going after too much at one time. For a program-driven DG effort, such as one under an MDM umbrella, the benefit comes from beginning to grasp the impact and interactions with the sponsoring effort.

Interestingly, you will start to notice areas where you are not sure what to plan for. This is how any required assessments are determined.

Ramification and benefits

DG is a business program. The business will benefit from this activity by getting a sense of how large (or small) the DG program will be. On one hand, a realistic image of the extent of the effort becomes evident. On the other hand, this realistic image may create some positive debate on where DG can work.

Sample output

See Figure 4.2.

Tips for success

Remember that you can be developing a rather large plan. Even if you are constrained to managing DG only for a specific effort, there could be a lot of moving parts to coordinate. That is why we had some other tasks to talk about capabilities and business areas. Sometimes a low-profile effort starts with good intentions, then finds out there are still too many interfaces to make it noninvasive. Then you need to go through a formal effort to identify the right kind of use case. Lastly, if time is of the essence, develop detail plans for the next few months and then add details as necessary, as long as you keep a good 3-month detailed tactical plan at the ready.

Approve scope and constraints

This set of tasks will present the desired team to leadership and get the support to move forward. The steering body must acknowledge the team's commitment with their day-to-day work areas. They then need to publicly voice support so there are no resources pulled back.

Approach considerations

Make sure you are getting true approval based on the scope and approach. Too often we see a "rubber stamp" and then shock and surprise when the leadership team starts to see resistance.

Lower profile efforts may not need a lot of activity around approval, but it is still good to reconfirm things with an overview of the approach to whatever sponsor the low-profile effort is leveraging. This is why we start to kick around what areas and capabilities can be involved earlier. Some vetting of scope should have occurred by now. The approval process should contain a good walk-through of the process.

Ramification and benefits

Obviously, approval to proceed with management consensus is a good thing. Additionally, the DG team now has, albeit modestly, the ability to say they are real and have some authority.

Make sure you are getting true approval based on the scope and approach. Too often we see a "rubber stamp" and then shock and surprise when the leadership teams starts to see resistance. The approval process should contain a good walk-through of the process.

Sample output

There is a sample deployment plan in the appendices.

Tips for success

The concept of "authority" is critical here. As long as the DG team has some perceived authority, it can proceed. If you have been staffed with marginal personnel, you can train them. If your steering body is weak, then develop tactics that will require more frequent vetting of progress and DG artifacts so you can continue to be noticed (in a diplomatic way).

Assessment

The Assess step gathers data about the organization's ability to do governance, and to be governed. These assessments can overlap greatly with other assessments done in conjunction with data quality, MDM, business intelligence (BI), artificial intelligence (AI), or other data solutions. They can also be derived as subsets of overall enterprise data assessments. The three types of assessments we will cover identify the perceptions and means an organization deploys to use data, and how the organization is positioned to carry out its day-to-day work while adopting a philosophy of managing data assets. The current state of an organization's information abilities, maturity, and content effectiveness is also examined. Your intent here is to not only confirm current state, but also discover obstacles to success and sustaining engagement. You actually start managing the upcoming changes for DG in the assessments. These assessments will supply relevant input to your sustaining activities later.

While we can get what we need from other assessments that are often happening around DG, this section focuses on them solely from a DG perspective. However, since there is overlap, a great deal of this chapter is similar in tone and content to Chapter 19 of *Making EIM Work for Business* (Morgan Kaufman, 2010). To see more assessment examples, as well as these assessments in a larger context, please refer to that work.

In the context of the DG assessments, the bottom line is that you need to understand if the organization is positioned to manage data or information as an asset. *Asset thinking* creates the philosophical basis for DG. Therefore, the philosophy must be accepted currently, or you need to start identifying the gaps that are preventing formal management of information assets from being adopted.

Assessments are more than just lists of questions that are asked in a stream of interviews.[3] They need to present an accurate, verifiable account of current state—and they need to do it in a timely fashion. Interviews can certainly accomplish this but are rarely timely. Therefore, the assessment for DG tends to be more accomplished via survey or other data-gathering techniques. You are doing discovery. Other forensic techniques are required and preferred over interviews.

---

[3] Along time ago I had a brief, mild, and short-lived reputation as a DM radical from a presentation entitled "Interviews are Dumb." It got people's attention. Consultants got panicky since the default starting position for anything seemingly MUST be an interview (it is not). Business personnel in the audience went "PHEW … thanks!"

All assessments need to cover the following dimensions (Fig. 7.6):

1. *Organization*—There are many aspects of the organization itself that will affect DG and, in turn, be affected by DG. This covers organization charts, distribution of staff, maturity of personnel related to information usage, and level of understanding of their data assets. It also covers the need for the basic skills required to exist in a governed information world.
2. *Alignment*—This dimension addresses, foremost, the business alignment to the actual current state of IT and information use. Are IT projects done within a managed portfolio, and is information a key consideration? A very significant reason that DG efforts go awry is a lack of business alignment to information. Without alignment, there is no business visibility to what a business program is. It gets lost in the noise and dies.
3. *Operations*—This dimension looks at the facilities that create and contain content. Whereas technology looks at the wire and pliers, operations looks at the usage of technology. Does the organization have operational processes and facilities in place to handle content efficiently? Are applications and systems process-heavy or data oriented?
4. *Technology*—Does current technology adequately support information use and creation?
5. *Data and Information*—What is managed in terms of information? Are privacy and security a concern? Are data ethics an issue? Are there rules and models from regulators or outside entities to be addressed?

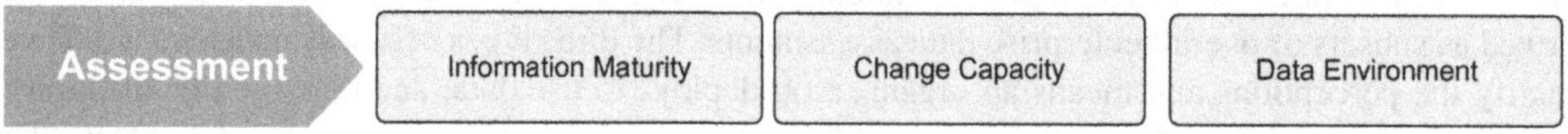

FIG. 7.6

Assessment activity

Information maturity

Information management maturity (IMM) of an organization may seem like a driver for DG versus a characteristic of DG. After all, if we were "mature" we would not need DG. It is a bit more involved than that. Anecdotal and hard evidence leads to the conclusion that organizations with a more proactive approach to information achieve better results. More and more companies are monetizing data or exploiting data assets as a standard business capability.

Maturity is not an indicator of if you "should" or "should not" embark on DG. It is not a report card that says you are "good or bad." It is a powerful means to articulate where you are in relation to where you need to go.

There is no relationship between maturity and ability to embark on governance. Research has shown that there is no connection between data maturity and the ability to implement a DG program. Granted, the details are different, but all that is needed is a firm commitment to moving yourself forward.

The key aspect of any discussion around maturity is that IT organizations, at a grass roots level, are beginning to see that there is a predictable maturity curve to climb around information production and usage. This, in turn, influences the definition of what the intended level of information maturity needs to be. There are definite stages of IMM along the way that can be described and measured.

While we review *how* the organization produces information and content, the main objective of this activity is to understand *what* the organization does with the content and information it produces. Usually this assessment is performed online over the company intranet. Questions focus on the relative impression management has regarding how well the company uses and manages data to its advantage. This includes use for decisions, communication, and analysis, as well as critical functions such as R&D or compliance when required by the business. Fig. 7.7 in the sample output section provides a brief sample of some of the survey questions.

There are formal maturity assessments that can be purchased. If you are in a financial services company, the DCAM™ assessment is available. Other companies can use the CMMI™ assessment. You can spend a little or a lot of money. A low-profile effort does not usually need a formal IMM survey and for sure not an external service.

However, if you are driving DG from compliance, or are pursuing advanced analytics and artificial intelligence, the first metric you pursue should be tied to a maturity level.[4]

Helpful hint

Sometimes a simple assessment is enough. After all, you would not be reading this book if you didn't believe the current state of data required some additional attention. So, an external assessment to find out what you already know can be a waste of money. Some of the external assessments are expensive. Make sure you are going to use the maturity score as a metric and commit your program to achieving capabilities that indicate a higher level of maturity.

Most frustrating are organizations whose executives insist on a really expensive outside assessment to confirm what the rank and file already know—the data is garbage and your DM capabilities are in the stone age.

You don't need the assessment to see if you need DG or DM. You need the assessment to see how far you need to improve things.

Approach considerations

Most likely, the length of the survey will be of concern to your sponsor or initial DG leadership team. A sponsoring CIO will be concerned with alienating stakeholders or ruffling feathers. Determining the scope of the instrument will be a function of determining what data you must collect for IMM, and whether or not you are combining this survey with another. The survey should take no more than 15 min in its online form or response rates will be too low to use. I should note that the formal outside surveys, like DCAM™ and CMMI™ are significant efforts. Many organizations do these as a means to sell management on the idea that the organization is deficient and needs to change. This type of survey is before any sort of DG program gets started. If these are done, of course, this activity is complete. Certainly, leverage the results. However, it is good to take some formal, short form of IMM if no other exists as a form of knowing where the organization is, and lining that up with the capacity to change.

---

[4]The two most common external assessment organizations are CMMI—https://cmmiinstitute.com/dmm and the EDM Council—https://edmcouncil.org/.

The actual questions need to be very unambiguous. A significant portion of responders will try to second-guess the survey. Always throw in a few questions that have obvious answers to indicate possible attempts to influence the results.

Of course, you want as many responses as possible. Respondents should represent, at minimum, middle, and upper layers of management. We prefer to segregate the responses of various groups, as their answers are almost always very different. In addition, there must be a mechanism to provide incentive to take the survey, as well as monitoring and follow-up processes to deal with laggards.

If the assessment is being done via facilitation or interviews, attempt to make the meeting as structured as possible. A group session should fill out the survey via a form, then tally and review the results. Interviews should cover a core set of questions in a survey format. The interview should also be used to collect personal impressions from interviewees. The population for interviews will be much smaller, so make sure the sponsor understands the IMM survey via interviews will be more anecdotal than statistical.

Unless you are low-profile and will get to IMM after an initial use case is completed, this activity is not considered optional, although it can be merged in with a change readiness assessment.

Business benefits and ramifications

This activity provides an objective view of the level of sophistication regarding information use. Often this survey will stand on its own to make a framing statement for the need for DG and can be used in conjunction with initiating second or third phase low-profile efforts.

Sample output

Fig. 7.7 shows a sample of some of the types of questions asked. All the surveys we use take this form of answer scale (called a "Likert scale"), which provides a decent distribution regarding the answers and gets us closer to seeing how the organization really feels about how data and content are used.

	Strongly disagree 1	Disagree 2	Neutral or undecided 3	Agree 4	Strongly agree 5
The enterprise has published principles on how we will view and handle data and information					
There are standards for how data is presented to all users, and standards within IT for describing data					
There are policies for managing data that are published					
The data policies are understood and adhered to consistently					
There are rules for sharing and moving data in and out of the organization					
There is a widespread understanding of the importance of data quality					

FIG. 7.7

Sample maturity questions

Fig. 7.8 shows two panels from the IMM results from the case study in *Making EIM Work for Business.* (UIC is the fictional company.) The maturity scale ended up as a 1.8 (subjective based on concurrence with sponsor and executives).

Question #	Percent Positive	Survey Question
26	49%	I understand the key indicators that measure my organization's performance.
5	72%	There are rules for sharing and moving data in and out of the company.
29	79%	I use data analysis to make changes in my work processes to improve results.
21	85%	My department has several databases, spreadsheets, or other data stores that we build and use to do reports.
28	94%	I collect and analyze information related to my work.

- There is general belief that management understands the measures of organizational performance.
- Given the insurance regulatory environment, the strong positive response to question 5 is not surprising; however it conflicts with general perceptions regarding data quality and controls.
- UIC management generally believes that it uses analysis to analyze and improve work processes.
- The high percent positive score for question 21, 29, and 28 that pervasive "shadow IT" may be exposing UIC to risks or higher costs.
- Question 28 indicates that most of middle management could be spending more time collecting and analyzing data than managing, and requires further review.

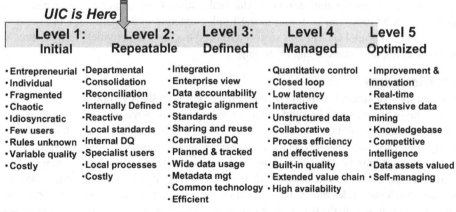

FIG. 7.8

Maturity survey sample results

Tips for success

Surveys have become a very popular means within organizations to measure just about everything. As a result, any attempt to survey may be met with suspicion or people may feel that they are not worth the investment in time. Depending on how survey results have been used in the past, you may be surprised at how far you need to go to convince personnel they will remain anonymous. If the survey history in your organization makes it a poor choice for you, consider facilitated focus groups conducted by individuals outside the DG organization. It will take longer but may lead to better results.

Time frames for this activity should average 2–4 weeks with the attention of a full-time resource from the DG team and assistance from an internal survey group. A short time frame is a success factor here. If there is a need to do focus groups, then assign two resources and get the groups processed within a month. Avoid the perception of "analysis paralysis." Remember there are people out there who will be looking for symptoms of the "same old IM project."

Helpful hint

Try your marketing or HR organizations for help. They usually do all kinds of surveys and are adept with them, and they can help you with focus groups, if you go that route.

Change capacity

All organizations are unique in how they carry out their mission and activities, even within the same business arena or market. This set of behavior patterns or style of an organization represents its culture. Part of any culture is its capacity for change. Obviously, organizations vary as to how easily or rapidly they can accommodate changes. Therefore, the objective of this activity is to measure this capacity for change and locate potential resistance points. If you do not do this, you risk missing vital information that will allow the team to accommodate and leverage your culture, rather than fight it. Cultures are not changed, they are leveraged. In addition, the earlier the cultural issues are identified, the sooner any large obstacles will be recognized and addressed. This is the start of organization change management (OCM). Additional OCM tasks can be found in the Implementation and Operating work areas. Very high profile efforts need parallel OCM plans so you will need to combine activities from all three work areas—Engagement, Implementation, and Operating.

Approach considerations

The Change Capacity assessment is strongly recommended. There really is no optional path—it must be done. It can be done in two passes, a brief informal iteration now and then a detailed formal pass during the Roadmap or Sustaining activities. The most common approach is to do a survey that is geared to reveal any glaring issues now, and then revisit the change capacity assessment during the implementation.

The low-profile effort may be able to sneak by without some sort of assessment of change capacity. Since many are considered noninvasive, in theory, there is no potential of a change issue. The team needs to carefully examine this, however. Moreover, if you are looking at working on a second phase, or expansion of DG, you may have been ok with the first pass, but as DG expands, you inevitably will become "invasive" to some degree. While researching this book, I had a talk with Robert Seiner, author of *Noninvasive Data Governance*. Noninvasive approaches are obviously popular, but at some point, someone will feel that there is a change and will be concerned. That is because, again, DG is an enterprise concept.

It is vital to assess the risk to DG that will originate from culture issues. The DG program must be sustainable and cannot be made so without vital information that will allow the DG team to accommodate and leverage the organization's culture. The results are used to adjust the Sustaining phase and will even influence the rollout of information projects and policies.

Some organizations will resist any assessment of culture from any sort of "technical" team. If the DG team cannot overcome this obstacle, bury the most telling aspects of the change-capacity instrument in another survey, such as the maturity or current state assessment survey.

For larger efforts and scopes, the target audience is all management, as well as knowledge workers or departmental analysts. At minimum, all stakeholders and their constituents need to be considered candidates for assessments. The population to be surveyed needs to be segregated with results kept by whatever segments you choose. At minimum, segregate upper management, middle management, and all others.

This assessment is in the form of a survey and is best done online. If an online survey option is not available, do focus groups. Given historically low response rates, the last resort is a form to be filled out. If the focus group or paper form options are used, allow several weeks to get focus groups scheduled. Then allow 2 weeks for forms to come back but expect 3 weeks during which they actually keep showing up.

If there is a hint of sweeping changes, or known resistance areas are already identified (i.e., a prior attempt at information management failed in some way due to resistance), then a formal instrument is strongly recommended.

Tips for success

Often business or technology executives that have not engaged in formal business change programs will resist performing this step. In fact, a lot of the Roadmap and Sustain activities are spent dealing with resistance. It will seem "squishy." However, any root-cause analysis of the failures of large technology efforts over the decades shows the reoccurrence of a number of significant factors—poor communication, no alignment with the business on what is to be delivered, ineffective training, and lack of business sponsorship, to name a few. These change-management issues have cost organizations millions of dollars in failed programs. If you want to do better with your DG program, you *must* formally manage the changes required.

The standard change management tasks used to support implementation of DG can be taken from any number of prominent organizational development industry sources or authors, including Prosci, Change Guides, LLC; John Kotter, or William Bridges. Please see the footnotes and bibliography for these sources. There is an *enormous* amount of material available for very little (if any) cost, and it is easily adapted to DG.

Ramification and benefits

The data collected from this step will be used throughout the program design and for a long time after rollout. It provides an excellent baseline to measure DG adoption as well.

Sample output

A simple visual is the best means to present results. Fig. 7.9 shows a strong, but not insurmountable, resistance to change.

Data environment

This third survey assesses the current state of people, process, technology, and data within the organization. The result of this activity is a data landscape/inventory, an understanding of the readiness of

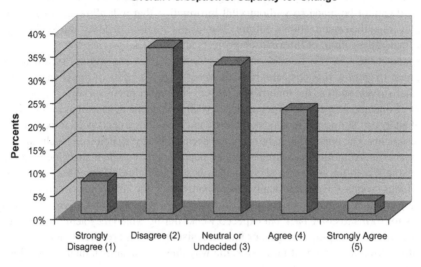

FIG. 7.9

Change capacity sample

the human element, the current deficiency and capacity of technology, and the existing mechanisms in place currently that handle and manage data. This survey can be used as a readiness assessment if done against some sort of benchmark.

People

The readiness of the human element is obviously important. But it is readiness for DG. The change capacity work will cover cultural aspects. This assessment will get into skill sets, who and how staff engages with data assets, and resource capacity.

Process

All organizations manage data, albeit many do it poorly. The reason to do DG is to manage data better. Understanding what you do with data in the current state is important, primarily so you know how far the journey will be to a future state. Most organizations have islands of success, or points of light, where data has been handled well and creates value. This part of this assessment will identify good processes and poor processes for data handling. Again, it can also serve as a readiness assessment if there are benchmarks available.

Technology

A careful consideration of how technology is used to manage and use data in the current state is valuable. At this point in the evolution of IT, most organizations have some sort of DM technology, as well as other technology that can be leveraged for DG.

Data

Knowing what data is used, and where it is, and how it is currently handled is key in discerning the gaps between current and desired future states. An inventory of all data sets, along with locations and metadata, would be wonderful at this point. Or, just the data assets that fall under current tactical scope and future scope. Many organizations are stunned to see the results from a data assessment when they see what is commonly known as the spaghetti diagram—with the hundreds of interfaces and data sources portrayed.

Approach considerations

This survey can be skipped for low profile efforts. Usually the environment is well known. However, as soon as you move beyond a tidy use case, you will need to have an idea where the organization is starting. Obviously, any kind of plan or roadmap is hard to do without a starting point.

More importantly, please be clear what areas (people, process, etc.) you will be assessing, and to what level of detail. For example, a really large, high visibility program may require skills assessments, and detailed understanding of current data processes.

A full data landscape assessment will include an inventory of data sources, external data sets, interfaces, downloaded files (like .csv), and Access databases. This can take a long time so plan carefully and do this only if needed to move forward.

Lastly, once in a while a client has asked me to look at the costs of DM. This consists of the labor cost of everyone using, moving, and maintaining data (not just IT—but all of shadow IT as well), plus license costs, external data fees, and infrastructure.

Ramification and benefits

Understanding the data environment is critical for understanding how to get DG operational. Most DG programs will eventually require some sort of data lineage or provenance, that is, tracking where data starts, goes, is used, and who used it. Very often a DG program will show immediate value when an understanding of the data environment is presented to company risk management. You are assembling the type of material that regulators will continue to request and insist on in greater detail as time moves on.

If you do look into the cost of ownership of data, I can guarantee that the total amount spent will be surprising, and there is a good chance that management will tell you they do not believe the number. However, it is quite common for the total cost of data to be four to five times higher than thought.

Sample output

A sample assessment is in the appendices.

Vision and plan

Few things are harder to put up with than the annoyance of a good example.
Mark Twain

The Vision tasks show stakeholders and leadership what DG will look like. This activity is the opportunity to "sink the hook" and ensure leadership is engaged. In some cases, a business case is required. This activity starts to lay the foundation for the business case and the final approach.

This means a bit more than a one-page picture, although that is important, too. This phase also includes work on the mission statement as well, and both are defined in detail below. A vision establishes a picture of where an organization aspires to be at a certain point in time in the future. The mission talks about what gets delivered to get there. The goal is to convey understanding and comprehension of what DG means and what the organization wants to do to get there. This vision reinforces the fact that the business of enterprise information asset management *is* the business.

Obvious business benefits and capabilities are presented to reinforce the "what" as well as promote the placement of DG as a business capability.

Vision can be an abused term. It implies fluff and waste to many disillusioned executives and staff. With DG, however, there is a profound need to convey the "big picture." Earlier we mentioned the need for organizational change management. A key aspect of a change program is maintaining a future vision in front of those undergoing the changes. Change does not happen among humans without some view of the big picture. What will a "day in the life" look like when DG is activated? What will be visible? What will be different? What business goals will be more achievable?

These activities are simple and should not take long, but we caution you to avoid defining a stated period for *fulfilling* the vision. Although nice, it is unrealistic at this juncture and, frankly, could alienate middle management who perceive such statements as arbitrary. In fact, this phase can result in a stalled effort if ignored or done poorly (Fig. 7.10).

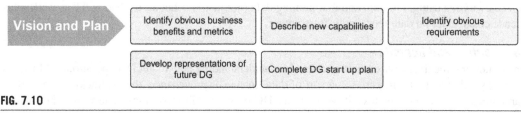

FIG. 7.10

Vision and Plan activity

Identify obvious business benefits and metrics

The organization's business needs can be translated to business benefits. If DG is going to support another effort, then you may have a lot of these benefits already.

The DG team needs to sit down and review any prior effort at business alignment or building a business case. A thorough understanding of what the business defines as success is essential. The team can then determine where the benefits are that have been identified and start to discover points where DG can affect or ensure usefulness of data and content.

If there are not any prior data/business interactions, such as advanced analytics, then you need to start to look at business plans and strategies. Either way, obvious areas of business needs will jump out—for example using analytics to create new products, or an MDM to provide a comprehensive view of customers. AI is a good example where business value is (or should be) tied directly to a data project. The results from this review should be listed and connected directly to obvious capabilities, which is the next set of activities in this section.

Business benefits and ramifications

There are benefits from examining benefits. If the DG team is not tuned into business needs, it will start to see where they can clearly speak to the value of DG. How can DG be a business program if it is not tuned into business direction? I am still stunned at how many employees in all varieties of large and/or well-known companies have no idea where the company is supposed to be headed.[5]

---

[5]While it is poor form to do so, I often want to take executives aside and educate them to the idea that informing employees of business goals is a good thing. They have a "need-to-know" or "elitist" mentality that assumes the person in the mailroom is incapable of understanding the goals of the company. While we certainly understand the need to keep *strategy* close to the vest, sharing the bigger picture with employees is an incredible contributor to an effective culture. (Everyone else will read about the strategy on the internet anyway.)

Besides the ultimate assignment of some financial value to DG, the business gets the material to think of DG as a business program vs an annoying IT effort.

Approach considerations

This is not a "read-it-on-the-train" activity. It is a "team-around-the-table" activity. There should be a summarization and presentation of the business programs that may require the application of DG to their information underpinnings.

Sample output

Fig. 7.11 shows an initial list and assignment of information management and therefore, DG needs. The columns represent a typical hierarchy of business alignment:

FARFEL EMPORIUMS SUMMARIZED GOALS			
Driver	Goal	Documented Objectives	Measurable Attributes
Improve Market Share	Recover lead market share in category	Regain market share of 25%	Market share
	Increase top line sales across all categories and stores	Increase same store sales 15% over three years	Same store sales, Forecasted versus actual
Increase Customer Interactions	Improve Customer experience	Increase visits per store from 3-4 per year	Store visits, Market basket return
		Improve service environment, highlight differences	Surveyed opinions
	Improve effectiveness of web site	Improve web sites sales 15% without cannibalizing store sales	Percent sales from website
		Integrate store and web site offerings	Frequency of assortment refresh. New products per season
	Increase repeat visits with more household awareness	Capture customer feedback, integrate findings into marketing	Store traffic
Product Innovation	Offer an improved selection of products and services by channel	Beat competitors to market on new products	Time to market averages for specified product
		Implement the most appropriate and profitable product mix, with brand consistency and neighborhood variation	Same stores sales, Category product turns
	Maintain accurate merchandising processes	Improve procurement and store communications	Elimination of missed products or out of stocks
	Improve R&D to improve recognize new opportunities	Gain insight into Generation x, y buying patterns	Demographic, psychographic trends
Improve Operational Efficiency	Improve management of merchandise inventory assets.	Reduce weeks of supply across appropriate product classes	Weeks of Supply
		Improve cash flow and asset management to improve current ratios	Current ratio, Inventory turns and Weeks of supply
	Identify business processes that can be improved to increase profits.	Improve processes through more efficient collaboration	SG&A expenses, Cycle times, Division results
	Optimize store performance	Monitor and assist stores with declining performance of more than 3% gross sales decline	Same store sales, Geographic and demographic sales potential
		Eliminate/relocate bottom 5% of stores	Same store sales, Geographic and demographic sales potential
	Reduce SG&A	Reduce "shadow IT" to competitive or industry standard levels	Total cost of ownership for data usage, IT and business areas

FIG. 7.11

Alignment of strategy to data and metrics example

- *Drivers* are industry- or market-inspired trends, usually stated in terms of a direction. These tend to be categories of business goals.
- *Business goals* are refinements of drivers, expressing the general trend in terms that indicate desired accomplishments within a time frame.
- *Objectives* are the specified, measurable criteria for achieving the goals and drivers.
- *Measurable Attributes* are aspects of the business to be measured. These become metrics or a category of metrics. Interestingly, you can also see that the measurable attributes are items requiring control—i.e., they are governed.

Tips for success

There are occasions when DG is initiated as part of a project with very good business justification and alignment, but the team is not privy to that information (see the footnote diatribe again).

This is not a showstopper. It means you need to add another businessperson to the team—one who can speak to business needs. In addition, do not hesitate to have the DG team sign an internal confidentiality agreement if the business goals are such a super-secret deal. If there is a total roadblock, then look at similar organizations, industry standards, or the *Wall Street Journal*. Remember you are just working on Vision at this point.

Describe new capabilities

When DG first started to become a serious element of the business world, we would match requirements to some sort of process model. For example, a company would start an MDM program. A declaration that DG was required would be issued, then some policies and resulting processes would be designed. Experience has taught DG practitioners that a better technique is in order.

Capability-based modeling and architecture are techniques initially used by enterprise architects. The value proposition of capability-based planning is the reason it fits well into the DG and DM areas. It allows for explicit alignment to strategy and promotes agile thinking.

A business capability contains more than just a process, or people perspective; it includes the data and physical perspective. In one pass, a capability represents process, technology, data, and people. For example, a capability of Project Support would encompass not just the DG resources helping IT but also the technology involved, such as a glossary or work flow tool.

Approach considerations

Chapters 8 and 9 will get into more details, but for now let's keep things simple around capabilities. There is not much new in the way of DG capabilities, so I have listed them for you (Fig. 7.12). Examining the initial business needs and benefits will point to some pretty obvious capabilities. That is all you need at this time. Remember, we are only doing a Vision.

Data governance strategy

DG strategy

- EIM and DG business alignment
- EIM and DG goal setting
- Compliance and privacy strategy
- Data ethics strategy
- Data principles
- DG technology strategy
- Business strategy support
- Data strategy support
- Applications strategy support
- Enterprise architecture support
- IMM/CMM strategy

Data governance management

DG operations

- DG activity management
- Data policy management
- Data standards
- Metadata and glossary management
- Measurement
- Issue management
- Data access and user
- Content governance
- Application development governance
- Compliance related governance
- Security/privacy governance oversight
- Ethics oversight
- Issue management
- Leadership communication
- Data risk oversight
- DG audits and controls
- Methods and workflow oversight
- Policy administration

DG measurement

- Effectiveness and efficiency metrics
- Measurements of data quality and usability
- Business impact metrics
- Data debt
- Literacy and maturity targets

Data governance definition

DG requirements and design

- DG assessment
- Federation requirements
- DG scope and focus areas
- Capabilities definition
- DG roadmap modification
- Controls specification
- Compliance identification
- Enterprise risk management specifications
- Ethics and privacy definition
- Policies and standards development
- Organization business information requirements
- Collaboration and communication set up
- DG metrics
- Metadata and model specification
- Taxonomy and ontology specification
- Data lineage and provenance specification
- Data classifications specification
- Data controls specification
- Data sharing specification
- Data integration specification
- Data life cycle management specification

DG frameworks and operating models

- Operating framework definition
- Engagement model definition
- Accountability and responsibility structure
- Data literacy
- Roles and responsibilities
- Collaborative framework
- DG technology requirements

DG technology

- Metadata management
- Data mastering
- Data lifecycle management
- Data security
- Taxonomy ontology
- Data movement and integration
- Reference data
- Data quality
- Data modeling
- DBMS
- Collaboration and knowledge management
- Stewardship

Data governance operations

DG communication

- Communicated expectations and accomplishments
- Communicated data-related directives
- Communication events and artifacts

DG training

- Technology training
- Data literacy awareness training
- Stakeholder and operations training
- Formal orientation and onboarding

DG services

- Data sharing agreement services
- Data integration services
- Data quality support
- Data compliance and risk support
- Data lineage and provenance support
- Data ethics support
- DG technology operation
- DG technology delivery
- Data ethics support
- Data delivery support

Sustaining data governance

Sustaining DG

- Organization change requirements
- Org behavior changes
- Leadership alignment
- Stakeholders management
- Conflict and resistance remediation
- Change plan management
- OCM plan implementation
- DG metrics
- Training oversight
- Resource development
- Promotion and branding
- Community and collaboration development

FIG. 7.12

Data governance capabilities list

Ramification and benefits

Without examining required capabilities, you will move from DG objectives directly to some sort of functional or process model. This does not allow a big picture. So, it becomes much harder to find what increments are best. As part of the vision for DG, the capabilities listed will:

1. Link business goals to required capabilities. This avoids the randomization effect—where an initial increment of DG is plucked from somewhere because it feels good or greases a squeaky wheel.
2. Clarify the business importance of DG.
3. Delay the desire to dive into "solutioning."

Sample output

There are two way to present the initial capabilities. One—just list them. Second (and preferably), show some sort of initial tie with business benefits (Fig. 7.13). It all depends on the type of program you are starting, and obviously a larger footprint effort would benefit from some sort of alignment.

FIG. 7.13

Tie business benefits to data capabilities

Tips for success

A couple of REALLY REALLY bad things can happen early in a DG effort and they stem from diving into the weeds. First, business hammers on "show value right away." This forces the team to think they need to get right into designing something, appointing stewards, and launching a bunch of stuff. Any flexibility or agility is lost. Second, in an effort to be productive, the DG team runs out and buys tools and consultants, and hires people. This all originates from failing to take a few moments to examine the capabilities needed and keep things lined up. The consideration of capabilities takes at most a few days for a large program, and a few hours for a low impact effort.[6]

---

[6]After 10 years of hammering on organizations to not do these dumb things, I hope that the capability-based technique prevents, or limits, dumb things. However, I am never surprised at the lengths people will go to dive into a comfortable place, surrounded by details that may or may not be of benefit or even the right thing to do. As soon as a new technology is floated, someone says "hold my beer" and dives in. This is a maturity deficiency but not in business resources. It is a lack of business acumen within DM and technology resources.

Draft preliminary DG requirements

This activity goes hand in hand with definition, capabilities, and an elevator speech. The identification of high-level data requirements allows you to organize a first cut at what is going to be governed. This activity starts with the business needs. In addition, known issues, requests, and works in process can be factored in to create a view of what is going to be governed. The team will need to consider specific business events, application maintenance requests, and regulatory or compliance obligations. From all of this, you will be able to point out major, obvious areas where DG can be applied.

If MDM, BI, or data quality efforts are driving the rollout of DG, it is likely that much of this information will be available. If you are getting into advanced analytics or AI, there is liable to be an attitude of "what requirements?" This is an oversight—there is still a business purpose behind the investment or desire to dive into advanced analytics. It may be "find ways we can monetize data to achieve market share" but that will suffice for now.

> A common challenge to new or rejuvenated DG programs is the "data governance does not apply here" syndrome. The Big Data and analytics explosion often featured data scientists doing a "collect, cleanse, and use" process. Many DG areas were told to stand back. However, the process for data science soon became "collect and pray" as a lot of models produced errors, and a lot of high hopes were dashed, along with the lack of returns on the investment. The same goes with building applications on service-based architectures. The technology is supposed to overcome any data issues. AI is another area like this and is downright scary—as machine learning has the capacity to intervene in daily lives. DG is a vital check and balance.

Remember DG is a program to ensure that Analytics, AI, BI, MDM, DQ, etc., all "stick." The drivers for these are drivers for DG as well. Interestingly, it is not until organizations place the DG alongside the other efforts (MDM, DQ, etc.) that they actually catalog and examine the business drivers of those very efforts. They are aware of the drivers in an anecdotal sense, but do not sit and catalog them. You might be able to sneak MDM or analytics in without a lot of consideration of business drivers (as long as you have a good sponsor and are solving a business problem). However, DG requires consideration of the business drivers and documentation of the following:

- How lack of DG-induced discipline could affect project or program sustainability.
- How lack of DG-induced discipline could increase risk by affecting ability to comply with regulation, loss of reputation from inadvertent unethical data usage, loss of market share, or potential lawsuits.
- Where DG and DM efforts are creating new business capabilities.
- The effects and details of long-standing requests to "fix" data in major applications.[7]

---

[7]After several decades of doing this type of work, it is amazing that among all of the documented needs and issues, all organizations can be counted on to have at least the following two DG drivers. First is the eternally lasting request to fix an old operational applications database. This application is the one that is so old no one can actually risk touching the code—so they try to fix the data, but the request has always fallen off the priority list. Second is that every organization has its legacy "data dumpster"—the ancient database that the data warehouse (and the second-generation data warehouse) was supposed to replace. One individual who has sole knowledge on how to navigate the thing supports it, and managers lay awake at night when they realize she is going to retire. Our theory is that all of the people supporting the legacy data are, in fact, related, and originate from an ancient medieval secret society of data wizards.

If you are standing up DG as part of an MDM or similar program, and cannot find any business drivers, you will need to execute a business-alignment exercise. The MDM team is in a lot of trouble but does not realize it.

Business benefits and ramifications

Some resistance to DG is inevitable. This step helps by consolidating the reasons you need to do DG. In conjunction with the elevator speech, you are beginning to develop the compelling message that will be required for successful organizational change. Again, we are getting stakeholders locked in through a solid vision and plan, so you need to have your facts lined up as well.

Approach considerations

Take the time to consider business needs in a formal manner. Do not throw it all in a document and say, "there it is." Remember, DG is all about making sure that data assets are being managed. Examine all of the points where data touches what the business wants to accomplish. At a minimum, you should break major business needs into subject areas or content used. Obvious and important metrics (or KPIs) are other groups of data that are used for data mastering or compliance.

A low-profile effort has an easy time—you are looking for specific areas and the requirements will be obvious. In some cases, the use case is so obvious that this step can be combined with design of the data management and governance solutions.

Sample output

See Fig. 7.14.

Representation of Initial Data Governance Requirements

Business strategy	Driver / Lever /Objective	Information Assets	IM Artifact Governed	DG Touch Point
Increase Value	Increase customer store visits by 2 per year	Customer, Customer analysis, Store activity, Store analysis	Customer data model	Customer MDM project
			Workflows and standards for consistency	Data migration and data quality remediation
		Business results from Store activity - Busines Intelligence, reporting, Analytics	Strategy Map of information levers and business benefits	Ensure aligned business vision of reporting, BI and analytics
				Aligned BI Architecture blending significant business information requirements into a uniform presentation
			Dictionary of core metrics and KPIs	BI and reporting requirements and deveopment
			Defined framework to deliver information	BI users

FIG. 7.14

Example of preliminary data governance requirements

Tips for success

If there is a lack of transparency when trying to align business needs with DM pursuits, or the organization is the sort where IT is told to "do what you are told," and information has evolved into the inevitable rat's nest of applications and shadow IT, look to external sources. Most industries have trade journals. Many companies must publicly disclose intentions. Use this type of data to interpolate a business plan and extract your business needs and DG requirements from there.

Develop representations of future DG

This activity is where the team defines a clear representation of what that aspirational "day-in-the-life" will look like. The emphasis is again on simple and straightforward. A one-page picture is often the only product for smaller scoped iterations. For larger efforts, you may need to portray a few other artifacts to show how DG can potentially operate.

Business benefits and ramifications

Besides the obvious advantage of the organization being able to comprehend what DG means, there is also the beginning of seeing the specific areas where DG can clearly add value. A successful outcome of this step will have the stakeholder seeing themselves as a participant.

Approach considerations

If you are doing DG for a very visible effort or for a large organization with a wide-span scope, then consider some professional help for messaging, the picture, or even animated media.

Sample output

Fig. 7.15 is an example from a client (modified for privacy) to show where DG fits. A large company was doing a global DG effort and the big picture was very important. It had to say that:

1. There was executive-level direction to use data as a game changer.
2. DG was going to be applied to all layers of their strategies.

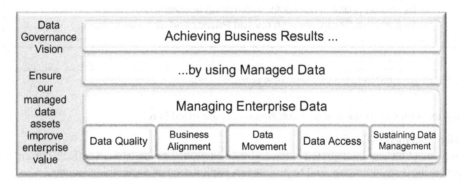

FIG. 7.15

Simple representation of what data governance does

Note the simplicity.

Fig. 7.16 demonstrates a much different example. A smaller financial services organization, well known for excellent service and execution, really needed to understand what happened day to day. Many graphical examples were produced, and none clicked. Again, simplicity won the day—a simple pro forma agenda for possible future meetings that shows what a "day-in-the-life" would look like. (Note that they referred to DG as IG—information governance.)

UBETCHA Financial Services Enterprise Information Governance		Thursday, November 21, 2010 2:30 – 4:00 Main Conference Room	
Meeting Topic:	Quarterly Information Governance Council Meeting	*Type of meeting:*	Udate, Issue resolution
Invitees:	Executive Data Strategy Council, Info. Governance Council		
AGENDA			

- Information Governance Value Update
  - IG and IM Scorecard Review
    - Data Quality Metrics
    - Business results from information projects
- Issue resolution
  - External business intelligence "cloud" package acquisition by Marketing
  - SAP Project - Location and Supplier Coding conflicts with current Ledger package
  - Marketing area absent from stewardship training
- IG Compliance items
  - Review IG Steward training progress (see issues)
  - Review status of BI and reporting governance
- Cross functional Collaboration
  - Report on recommended enterprise data controls
  - Report on policy revisions for information accuracy
  - Report from Compliance on revised Privacy and Security policies

FIG. 7.16

Example of data governance in action - sample agenda

Lastly, examples of mission and vision statements are listed here:

Mission Statement—Retail

To implement a shared, integrated enterprise data environment that always reflects current and future business requirements, and to promote its exploitation as a valuable resource.

Vision Statement—Retail

An organization positioned to act faster, more effectively and efficiently in a dynamic business environment due to a cost-effective data resource managed as an enterprise asset.

Mission Statement—Energy Company

Ensure that information management provides the resources, processes, and enabling technologies necessary to manage information as an asset throughout its life cycle.

Vision Statement—Energy Company

... will manage its information in a disciplined and coordinated manner to optimize the value of our investment in information assets, support effective and efficient operations, mitigate legal and regulatory risk, and improve the delivery of services to our customers and stakeholders.
Vision Statement—Insurance

... will use information management and governance to enable employees, customers, and business partners to have easy access to the information they need any time, any place, any format, reduce the cost, or improve the value of investments in information architecture, and ensure data accuracy, quality and consistency.

(I do not recommend the insurance example, as it was in existence when we got there. It is a little too focused on information technology, but it set a baseline for measuring and achieving the mission, so it sufficed.)

Tips for success

This activity is, fortunately, one where it is effective to gather many examples from elsewhere. Be certain that the definition contains elements that are meaningful to your organization. If discipline will be a challenge, mention it. If authority is important to success, mention it.

Ideally, your MDM, analytics, or other project will have mission and vision statements that can be leveraged. If not, you will need to assist in these efforts by creating them. Remember the fundamentals of *mission* and *vision* statements.

- A *vision* statement should provide a picture of where an organization would like to be at a certain point in time in the future. As such, it must clearly state what is to be accomplished, therefore supplying a foundation for measurement by framing the goals and objectives. For example, a vision of "we are going to be the best" is not very clear.
- A *mission* statement is a carefully worded statement of what an organization does in support of its vision, goals, and objectives. Each word of the mission statement is chosen for a specific reason.

The elevator speech needs to be positioned similarly. You also should fold in what DG will do for the company, using terms such as "ensure value, increase revenue," etc.

Whatever you do, *never* use words such as "better data," "improved decisions," or similar terms. They are vague and irrelevant to most executives.

Complete data governance start-up plan

This set of tasks develops the plan for the definition and roll out of the DG program. Any required constraints are applied and any results from the assessments are factored in to make adjustments. This activity can take place at any point within the Engagement set of activities, but usually it is best to get as many contributing factors understood before finalizing the approach.

Business benefits and ramifications

When the team starts the detailed planning, the full extent of required activity will become apparent. Scope can be adjusted here. This prevents the DG effort from possibly going after too much at one time.

Approach considerations

For a program-driven DG effort, such as one under an MDM umbrella, the benefit comes from beginning to grasp the impact and interactions with the sponsoring effort.

For low profile efforts, usually the assessments are minimal, so task planning can occur in parallel with other engagement efforts.

Should you need to apply DG to an existing effort, such as analytics or AI, then make very sure you coordinate your planning with the other efforts. Do not step to the sidelines. This is the point you need to insert DG into the other projects. You need to insist on the role as an enabler, not spectator.

Sample output

There is a sample DG project plan in the appendices.

Tips for success

Remember that you can be developing a rather large plan. Even if you are constrained to managing DG only for a specific effort, there are a lot of moving parts to coordinate. If time is of the essence, develop detail plans for the next few months and then add details as necessary, as long as you keep a good 3-month detailed tactical plan at the ready.

Engagement case study—Rocky Health

Since Rocky Health is a program that is underway, the Engagement and subsequent activities will reflect how to expand a low-profile, or low impact program, into a broader set of capabilities. It will also offer insights on what can be done when DG capabilities are minimally staffed.

Remember Rocky's approach needs to be low profile and they are not heavily staffed. While getting the organization fully engaged is important, we are not going to be able to go into a set of activities for a broad audience. So, the checklist is used by reviewing each activity, and developing a sense of how this effort will accomplish each item. The resulting checklist for this, and all the other areas, are in the appendices.

Tom is the DG lead for Rocky Health. As he and the consultant review the DG checklist, he notices that there are many activities in each topic of each work area. All of the topics seem mandatory. However, since Tom also knows that he must proceed with a lower profile, highly iterative effort, he needs to adjust how he approaches *each topic*. In addition, Tom needs to get a sense of how to get the most out of his consultant, as he has to maximize a limited budget.

Tom and the current sponsor perform the Engagement activity in about 7 business days—most of it focused on identifying stakeholders, then the next likely business areas, with the potential benefits. The Vision and Plan activities took place during a facilitated meeting, and Tom provided lunch so as not to impact normal business operations too severely. Tom finished the project plan based on the remaining work area checklists, the results of the facilitated session, the possible tasks, and review of the pace with the project sponsor. He then created two sprints using those tasks. Tom drafted a small set of metrics to enable progress tracking. Finally, two of the stakeholders were invited to sit on a temporary steering body to form the beginnings of a new future DG council.

Here in Fig. 7.17 is a sample vision from this work area.

Rocky Health vision

Fig. 7.18 shows how the checklist was used to create a specific approach for Rocky Health.

- DG at Rocky
  Health is not a new project or
  initiative. It is a change in
  behavior and approach to work
  already underway or planned

- Rocky Health DG will define a
  new way to work, not new
  work, and direct the new work
  at improving regional care

- Behavior changes in data and
  information handling is
  mandatory for success of DG

FIG. 7.17

Rocky Health Vision of Data Governance

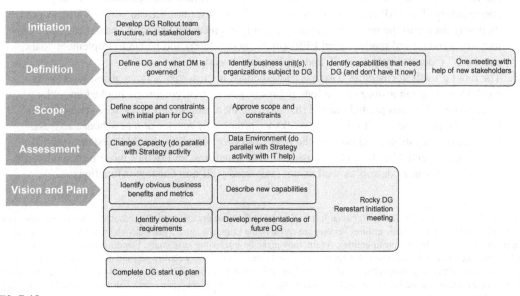

FIG. 7.18

Rocky Health check list example

Rocky Regional Energy Coop case study introduction

Rocky Regional Energy Coop (RREC) is an electric power organization in a large western state of the United States of America.[8] It serves a major portion of the state, with a mix of rural and city customers. It generates its own power from several hydroelectric and one coal-powered facility. It often needs to buy power from other grids, as its customer base is far outpacing its ability to build more generating capability.

RREC has recently initiated its second DG initiative. The first effort started 5 years prior within the CIO office. It was a combined data strategy/governance team, being stood up in response to the aftermath of a very painful SAP implementation. A large integrator did the implementation, and while great attention was paid to functionality and training, little or no attention was paid to data policy and use. Additionally, cultural issues were ignored, and often steam rolled by the integrator. Since the integrator is long gone, the remaining hostility is directed at IT. As one would expect, that hostility did not help the nascent DM and DG efforts. An enterprise information management (EIM) program was designed and featured a comprehensive data strategy. A few isolated DG teams were set up. Stewards were identified. Several training sessions were held, themed t-shirts were given out, and then everyone went straight back to doing things with data the same way they always had done them.

Some examples of the on-going struggles with data are:

1. Over 10,000 Access databases used to develop "standard" reports, including reports to regulators that oversee environmental issues and approve the rates that RREC can charge for energy. Recently federal agencies have begun to cross reference reports from utility companies and descended upon RREC with the "why don't these numbers agree?" questions.

2. Energy generation and distribution is engineering and capital intensive. Over the years the various disconnected systems affecting equipment and facilities became unbearably contradictory. The SAP effort fixed the issue for 6 months, until the old ways of DM worked their way back into the new applications. Recently engineering acquired assistance with a new industry standardized model called EPRI (more on that later) and began to implement "data standards" independently from IT and applications oversight.

3. The rapid growth in customer base has caused chaos in customer service areas. Recently a local TV station ran a story on the long delays in getting new customers hooked up to the grid. Some politicians publicly stated that any rate increases would be closely scrutinized in light of the poor service. It takes 3–4 weeks to initiate new service at a new house address. Root cause analysis pointed out that the new Customer Master in the SAP system was still being supplemented by three other customer data structures, and two of those turned out to be deficient in privacy controls as well as not integrated well into Customer functions.

---

[8] As with the other case study, Rocky Regional Energy Coop is a fictitious organization. It is based on several clients as well as energy industry trends. "Public utilities" as they are called in the United States are generally chartered as highly regulated, but publicly owned, often nonprofit entities. Again, they make an interesting case study, blending severe regulatory oversight from state and federal entities with a capital-intensive business, technology challenges, and deeply embedded cultures. Also, any resemblance to persons alive, dead, fictional, or real, is purely a coincidence. Very often I have had clients tell me "Hey—we were that case study in your book." Actually, there is never one client represented. I also had one tell me that they were glad the case study was not about them, as no one could possibly believe one organization was so messed up.

The new CEO has instructed that a Chief Data Officer (CDO) be appointed (not hired) and RREC needs to start to leverage its SAP investment along with another attempt at EIM to get the data assets in order.

Engagement case study—Rocky Regional Energy Coop

The first order of business for the new CDO was to get some external assistance to reenergize the DG program. However, given the prior history of a large EIM program dropped in by outsiders, it was determined to have the external help mentor, assist, and train the RREC staff.

From the beginning, the new CDO (Diana) understood that, while the culture of a utility company can be stodgy, doing things for small areas would never develop an enterprise sense of "data as an asset." In addition, the very nature of the work stream already in place (EPRI, enterprise resource planning data issues, and the connection problem) mandated a broader response to DM.

As mentioned, Diana was an internal placement, from a management position already at RREC. She was previously Director of Standards and Safety, being responsible for ensuring that safety standards were well documented and followed. In addition, her standards role also meant she was given the

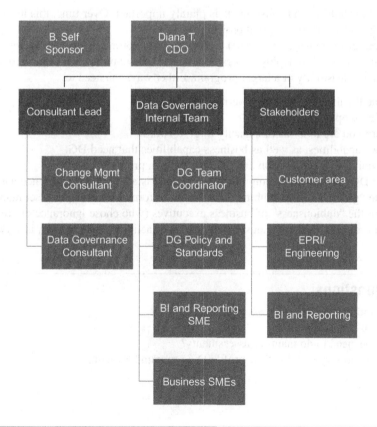

FIG. 7.19

Rocky Regional Energy data governance deployment team example responsibility several years ago to look into document management. Diana started her career at RREC in the AppDev area as a developer, project leader, then a manager, before moving into Standards and Safety. Diana and the consultant start to jointly plan, knowing that the Engagement work area will be crucial in getting an effective and recognized effort to a sustainable level.

Diana and the consultant started by assembling her team. She needed to get some key stakeholders on board as a steering group and recruited some individuals from business and IT areas that were familiar with standards and models. Unlike Rocky Health, Diana has several sessions with stakeholders to ensure understanding of affected area, nature of the program, and potential effects on current state capabilities. Obviously, there was not any deep awareness and support for DG. But remember the purpose is to get stakeholders engaged.

The overall time for engaging stakeholders and getting a vision in front of them along with an approach took about 3 weeks.

Fig. 7.19 shows the type of team Diana had to initially assemble.

Summary

Engaging your stakeholders and constituents is plainly important. Over time, this has become as significant to starting a DG effort as a good project plan.

Initiating your effort to build (or restart) a program does not mean you have developed a business case and have full approval to deploy. It means you have the support to get organized, start a smaller effort, or lay out the feasibility of a larger program. Either way, you need to:

1. Cleary define the meaning of data governance.
2. Clearly define scope.
3. Assess where you are in terms of capacity to go forward.
4. Identify new capabilities, as well as business capabilities that need DG.
5. Make sure you have a good vision and plan to deploy a program.
6. Framing the DG program in a comprehensible manner is a key step. Experience has shown that not everyone "gets it." If you combine the newness of formal information asset management with some of the "dubiousness" of business executives (who chose ignorance or are soured on information projects), then you can see that this set of activities, while short, is very important.

Essential questions

1. Why is it important to obtain engagement?
2. Why is buy-in "a possible problem"?
3. Do you always need to do maturity assessments?
4. Should you always try and do a low-profile DG program? Explain.

Strategy

Chapter Outline

> *Strategy without tactics is the slowest route to victory. Tactics without strategy*
> *is the noise before defeat.*
> **Sun Tzu**

Overview

This section is dedicated to strategy. The Strategy work area means there are a set of possible activities making sure data governance (DG) has value and meets business needs. You need to examine these, and then determine what is required for your particular situation. DG will always have multiple elements at play: people, psychology, organization goals, timing, conflicts, technology, and external factors. All of these elements need to be pointed at achieving several goals. This is the essence of strategy.

You will notice that some activities are similar to engagement activities. The purpose is different for items like capabilities and use case, as is level of detail. Remember, for a smaller, low profile effort, you may run the checklists and combine many of these activity areas. The activity areas exist to provide thought starters, not provide a recipe.

Data Governance. https://doi.org/10.1016/B978-0-12-815831-9.00008-4

We will see both of our case studies have to do some sort of strategic thinking, either to start enlarging the small program, or restart a large program. Regardless of the nature of the DG effort, there needs to be a plan to go forward. Activity from this area may take a few hours or a few weeks based on approach.

DG strategy requires the DG team, regardless of size, to address business alignment, defining the value of the program, and then an understanding of the high-level requirements that must be met. Obviously, from here we can feed into design, architecture, roadmaps, etc.

Fig. 8.1 show the types of topics in this work area.

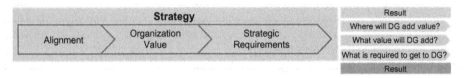

FIG. 8.1

Strategy work area

The "alignment" topic is critical—after poorly managed change, the number two reason to fail at DG is to not do anything to benefit the organization, in terms of revenue, savings, expansion, risk management, etc. I know they are tempting, but "foundational projects" like loading up a glossary before there is any defined business application are useless. I deliberately use that strong language, having tried them myself many years ago as a way to desperately show some sort of progress. Now we understand that the new business capability of DG needs a business justification and reason.

"Organizational value" is a separate topic because it creates the foundation to measure the contribution. Not necessarily justify the work, but contribute to its sustainability.

Lastly, we need to look at current and new requirements being placed on data. From there, "strategic requirements" are developed and examined to get a sense, even at a planning step, of how to parse the upcoming work into manageable pieces. Fig. 8.2 shows the activities within this area.

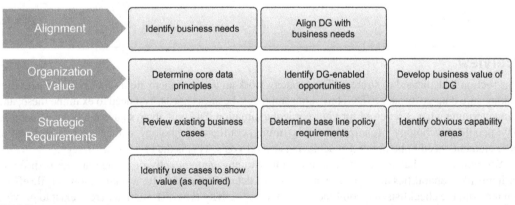

FIG. 8.2

Strategy work area activities

Alignment

Alignment refers to the direct linkage of the efforts to manage information assets to organization or business strategies and measuring these information and knowledge projects against the anticipated benefits.[1] This chapter will focus on the particulars regarding DG.

For example, an exercise to do the business alignment for a refit of a data warehouse would (or *should*) focus on the business benefits that will be derived from using the more accessible data or having powerful analytics. In the context of DG, we want to make sure that the business benefits are actually achieved (i.e., the right things are done to ensure successful use of the new data warehouse). This may seem like a faint distinction, but it is very important when your DG program is asked to justify its existence. Your business value for DG is stated through the same lens as the value of a Six Sigma program or a compliance program. DG is a necessary underpinning. Therefore, the DG rollout team needs to make sure that the link between managing data assets, organization strategy, and DG are apparent. This provides the baseline for sustaining the DG effort.

Identify business needs

Even the smallest effort should understand the possible impact of DG to the organization's goals. That means leaning about what the business needs. This is an objective exercise. It is not about discovering what kind of data business areas want. In fact, if you can access strategic plans or planning documents in general you don't even need to ask anyone.

If a 21st century organization has any kind of plan or has defined a set of targets, then someone, somewhere, is messing with data. So understanding business needs leads to an understanding of what might be happening with data, even if no one is coming right out and saying it. In fact, usually no one WILL actually come right out and talk about what the department data analyst is doing. This activity serves to familiarize the DG team with business needs and deepen the position of DG as a business program.

Approach considerations

This is not a "read-it-on-the-train" activity. It is a "team-around-the-table" activity. There should be a summarization and presentation of the business programs that may require the application of DG to their information underpinnings. For the low profile effort, allocate a small amount of time (a few hours at most) to talk to a business leader or review strategic plans. Gain an understanding of how your immediate efforts could be leveraged longer term. Figs. 8.4 and 8.5 recap business needs and alignment for the use cases.

For a program restart or a DG program that needs to cast a wide net, understanding where the organization wants to go and what data is needed to get there will start the foundation for an engaged roadmap. Plan on not only understanding but documenting the business goals and needs. This artifact is then used to map DG activity to business needs later.

---

[1]John Ladley, *Making EIM Work for Business* (Waltham, MA: Morgan Kaufmann, 2010).

Ramification and benefits

Regardless of your approach, the DG team should have a sense of how organization needs and goals are candidates for being supported in some way by DG. In addition, the financial aspects will be known in case you need to develop a formal business case.

Align data governance with business needs

Remember—DG needs to support the business. It is a business capability. It is really important. Often, I find a CIO (Chief Information Officer) to be a major obstacle to good alignment of business and DG. They claim to own technology, and data is technology, so they own governance.

Business alignment is mandatory if your organization wants to get more from data than simple operational support. If you are in a data intensive industry, such as healthcare, DG is now the determining factor of success vs mediocrity.

If you cannot state the outcome and operation of DG in a business context, then the simple, ugly fact is you may not succeed.

This activity serves to collect and analyze business objectives, goals, and drivers and in effect, create the foundation for a business case, whether one already exists or not. Of course, if you are engaging DG as part of Analytics or master data management (MDM), you may have a business case, so make use of it by all means.

The objectives here are to:

1. Derive the organization's goals and objectives and look for DG opportunities to support business objectives.
2. Develop sufficient business information to provide input for the DG team to determine some financial impacts of DG.

Most methodologies and analysis systems have a hierarchy of goals, objectives, etc. For this text, the hierarchy is as follows:

- *Drivers* are industry- or market-inspired trends, usually stated in terms of a direction. These tend to be categories of business goals.
- *Business goals* are refinements of drivers, expressing the general trend in terms that indicate desired accomplishments within a time frame.
- *Objectives* are the specified, measurable criteria for achieving the goals and drivers.
- *Measurable Attributes* are aspects of the business to be measured. These become metrics or a category of metrics.

An example of a driver is *customer intimacy*. A goal would be *improved customer retention*, and the objective would be *increase customer retention to 97% this year.*[2]

This activity is not as detailed as it could be if the team was doing a full-on enterprise information strategy. There is just enough analysis to show how DG can hold up its end in the business environment.

---

[2]Ibid.

Ramifications and benefits

If there are no obvious business cases tied to any information management-type solutions, then you need to do this. The most typical scenarios we see are:

- when a CIO starts an MDM project strictly as a technology effort;
- an advanced analytics areas set up by a business area, such as marketing;
- an enterprise resource planning (ERP) software package used to integrate data with no business drivers.

All three of these scenarios usually result in a revelation that the data is not up to the task and will be a barrier to success. There is no revelation—all three of these, and any other project or program that is data intensive, requires DG. You need to find the business benefits and get DG away from being a foundational project requirement.

This step captures, in business terms, how the enterprise needs information and content to achieve its objectives. The DG (and all of enterprise data management) can be tied back to this list, ensuring that information assets are used and managed to meet business needs. In Fig. 8.4, follow the sample from organization strategy through to the required DG items.

Approach considerations

This activity is presented here because all too often the CIO is told, or tells some subordinates to, "get some data governance running." When the typical scenarios presented above happen, the projects tend to leave a smoking crater. Then the realization sets in that DG should have been deployed as part of the Analytics, MDM, or ERP project. In these settings, there has been little business input, so the outcomes of DG cannot be defined even if the DG team manages to get it up and running.

Use existing documents or discovery techniques vs interviews if possible—look for business plans, external research, internal project return on investment (ROI), budget, and management by objective type documents.

If you are low profile, then make sure you first understand what immediate type of business benefit you need to deliver. That may be all you do for this activity. Or if you are restarting an initiative, this material may be available, and all you are doing is making sure you have the business context. So, this activity could only take a few minutes!

Even for larger efforts or restarting a larger program, this activity can take only a day or so if the business plans are readily available. The real issue is if you have nothing, or a prior version of DG was foundational only. It may then take between 1 and 2 weeks, depending on the span of DG in the organization. The most time-consuming efforts in this activity will be document research and review and building some sort of strategy map. Additionally, if interviews or reviews of findings are necessary, there is additional time required for scheduling sessions with management.

When you do not have access to business plans, then you need to take the "guerilla approach" to business directions. The DG team should look outside your company at the business environment. Mass media, trade publications, and regulatory filings are excellent sources of business-direction tidbits that can be used to support information asset management and DG.

"Business drivers" can also be a vague term. Most organizations have goals and objectives, or strategies. There are many layers to corporate and organizational strategies. Most list drivers as some sort of influencing factor. Therefore, be cautious of semantics. A great deal of the time, goals, objectives, and/or drivers can be discerned from corporate documents. The key is to look for material that lists or implies *measurable* goals and objectives. There are several fundamental reasons to take this perspective:

- The ability to share data is not of prime interest to a CEO—unless you have a specific data monetization effort, the CEO focus is lower costs and more revenue. Period. If sharing data leads to those, fine; if not, don't bother them.
- Most executives have been interviewed into a stupor. However, few DG or enterprise information management (EIM) teams know enough about the business to replace executive insight—so other techniques are called for.

Once the goals and objectives are discerned, then it is very important to consider what the enterprise is like with formal information asset management (IAM). Where does IAM fit into the current business? This means we need to start to recognize specific actions that the business will take. This may mean creating a list of specific activities where the business will use information to accomplish goals. If you are lacking any meaningful input from leadership, you may use this cheat sheet (see Fig. 8.3).

Usage Value Category	Data, Information and Content used to improve or achieve goals through:
Processes	Improve cycle time, lower cost, improve quality
Competitive Position	Capture competitive intelligence and differentiate yourself
Product	Create package and market unique, higher margin products
Asset/Intellectual Capital	Prolong leadership, embed knowledge into products and services
Enabler	Foster employee growth and empowerment
Risk	Manage risk, of various types, that threaten value by increasing liability

FIG. 8.3

Data usage for value, or monetization

It is a simple matter to take an objective and "bump it" against this list of six basic generic activities that take place around data and content to monetize data. Your goal is to see which of these basic uses of data can be applied to help achieve the business goal you have identified.

Sample output

See Fig. 8.4—it represents the summarized output for Rocky Energy from this activity, and you should derive the same type of work product. Fig. 8.5 is another example, however, in the form of a strategy map, which is now a more common presentation and is based on the approach by Nolan.[3] (It is the same type of document as Fig. 8.4, only flopped on its side.)

---

[3]Nolan, *Strategy Maps.*

Driver	Goal	Documented Objectives	Initiatives	Business Data Mgmt Needs	Possible DG Capabilities
Growth in operating margins	Increase Revenue	Increase nonenergy product revenue 15%	Increase sales of existing products and services		
			Introduce programs to encourage efficient consumption of electricity		
	Reduce Costs	Reduce tool and material redundancy 25%	**Minimizing the tools & equipment needed to operate our business**	**Tool and equipment data management**	**Data lineage, data quality**
		Improve power output 5% in existing plants	Minimize production costs through improved plant capacity		
			Improve "capital efficiency" (getting the most out of every capital dollar spent)		
	Maximize return on assets (plant, people, processes)	Improve power output 5% in existing plants	Improve availability of information to facilitate efficient operations		
			Increase load factor on system		
			Effectively evaluate business opportunities in approval process		
		Reduce new transformer installation cycles by one week	**Attain more efficiency in asset set-up and management**	**Tool and equipment data management**	**Data standards**
	Position for Growth	Improve engineering project results-80% on time and budget	Plan, acquire, and position new and existing assets and resources (distribution, transmission, coal, wind, distributed generation, etc.)		
Effective Regulatory Position	Improved image with regulators	Reduce customer complaints 25%	Monitor customer privacy		
			Provide strong analytical and fact-based, well-documented positions		
			Establish strong support of and representation in community		
			Meet reliability targets		
			Meet compliance standards (e.g., reliability standards, call center responsiveness mandates, environmental mandates)		
			Analyze and communicate action steps necessary to minimize reported customer complaints		
Customer Satisfaction	Increase Value to Customer	Reduce new connection time to one week	**Improve responsiveness (cycle time, on time, information, etc.)**	**New service appointments, scheduling and asset availability data**	**Item, inventory accuracy, data quality**
		Increase customer satisfaction on new service to 90%	**Deliver service to expectation levels**		
			Increase system reliability		
			Increase options to customers		
			Limit price increases		
			Increase value of offerings (maximize service received for each dollar spent by customer)		
			Improve Ability to Anticipate/React to System Swings		
			Ensure customer data is used appropriately and as specified		
			Reduce Interface Cycle Times		
Operational Excellence	Improve process efficiency (cost, cycle time) and effectiveness	Reduce cycle times and costs	Identify and define (map) processes to be able to effectively execute business strategies		
			Define process metrics with appropriate goals/targets and control limits		
			Provide tools/technology to effectively enable our processes		
Risk Management	Improve regulatory risk management skills	Reduce risk exposures by 8%	Ensure customer data privacy compliance		
			Meet compliance standards (e.g., SOX, Environmental, Employment, etc.)		

FIG. 8.4

Rocky Energy alignment

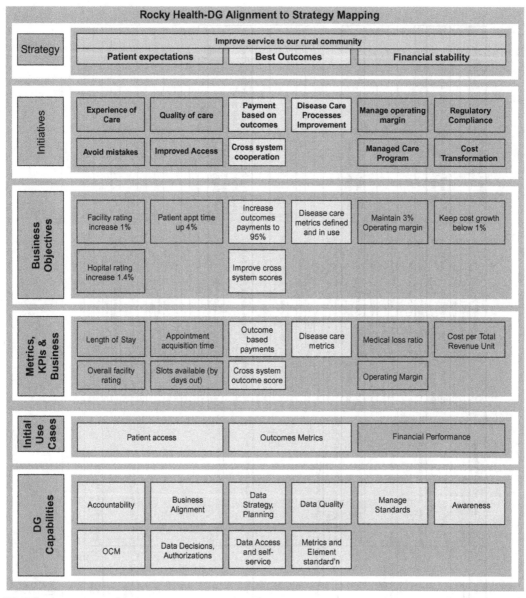

FIG. 8.5

Rocky Health alignment

You may be questioned as to how you reached certain conclusions in terms of business needs and direction. This is indicative of scarce or low-level sponsorship. Maintain all of your research so you have an audit trail. It does help when presenting your business findings and, hopefully, management will take note. The DG team is likely to get the attention of management since there will be the appearance of some serious strategy work going on—without management's involvement. Of course, the best critical success factor is to have executive sponsorship and presence as soon as possible.

Helpful hint

I did an alignment activity for a large company where the business goals were "not for our eyes," followed by "we just need a data strategy, so you don't need to know about the business strategy." As I mentioned above, we used publicly available documents to get a sense of where the business might be headed. My team did the research and data gathering required to find business needs that matched up to DG. We then developed a strategy map and presented it to the DG steering committee. The uproar was astonishing, and the entire team was interrogated immediately as to where we had found this "classified" information and who was responsible for "leaking" it. I revealed that the "top secret" sources were the *Wall Street Journal* and the CEO's letter in their own annual report.

However, the lesson learned was to vet the findings within this team and executive sponsorship first, and then drive to the final deliverable of this work area, which is an idea of the financial impact of DG. You may never get to show your wonderful work to anyone.

This activity may take a few lunchtime sessions with peers, but a mildly facilitated discussion will enable a small group to derive a list of 10–15 items. You will have collected sufficient business data if you have items that, if addressed, would reduce risk or drive good numbers to financial statements. There should also be items that reflect product or service changes, efficiency changes, or customer-relationship improvement.

When deriving the levers and examining how data will get the business where it needs to be, we facilitate a meeting where selected business personnel view each goal or objective, and look at each of the six usage categories in Fig. 8.3. Often we have them complete a phrase: "ACME Company will use data/content to as a means to achieve (insert the goal here)." The answer to the phrase is a specific opportunity for DG to support business value. For example, a completed version of the phrase might be: "ACME Hospital will use data/content to "target healthy lifestyle messages to members" as a means to achieve "higher member retention and lower health care costs"." Again, you are doing this for attaining a business context for DG. The ideal scenario is a room full of inspired and authoritative business leaders, but it may be a team exercise done solely for making sure DG can measure itself and be sustainable.

Organization value

Our principles are the springs of our actions. Our actions, the springs of our happiness or misery.
Too much care, therefore, cannot be taken in forming our principles.
Red Skelton

This set of activities is used to connect financial benefit to DG.

You cannot manage what you do not measure, and this phase gives you the ability to measure the success of managing DG. But you don't do this section for an ROI necessarily.

The real reason to do some kind of business benefit, whether you are low profile or not, is to defend DG. Regardless of initial perceptions of qualitative value, people being people, rapidly declare any program an inconvenience as soon as they need to set aside time to deal with a new policy. Unless you hold up a paper that says, "If you do not do this, it will cost us $$$$$$," your DG program is defenseless. Even if you have a strong sponsor, you need to keep a business case in your hip pocket to counter resistance and be sustainable.

This activity has changed since the first edition. I moved development of data principles into this activity area. Practical experience has shown that talking about business value and principles often appear in the same conversation. I moved the checklist discussion of principles alongside discussion of business value (Fig. 8.6).

FIG. 8.6

Organization value activity

Determine core data principles

Principles are statements of values and philosophical beliefs that the organization wants to adopt. There are a few ways to look at the benefits to be derived from development of the enterprise information principles.

1. The principles anchor the formation of policies. The core information principles expressed in this step will be used to frame the procedures required to enact DG. It is not hard to see there are some ramifications requiring policy and process if we examine the most common principle—that is, "we will treat information as an asset."
2. The defining and vetting of the new principles create a deeper layer of understanding of DG's meaning to the organization. DG concepts and impact come into sharper focus if leadership is still uncertain as to its meaning.
3. The review, refinement, and publication of the principles establish the relevance of the DG team.
4. Very often principles lead to an immediate perception of value; that is, if a principle eliminates debate, or rogue shadow IT, it is perceived as saving money.

This activity sets the tone and foundation for the entire DG program. Without review, participation, and buy-in, it is impossible to extract relevant, realistic processes and policies. If the DG team and the constituents of the DG program are not immersed in the principles, then there is no intellectual connection between policy and philosophy.

While it is good to start with a "seed" list of principles, which usually comes from external examples and existing internal belief statements, my own list is called GAIP™ as it presents a framework of principles. Honestly, the ones listed in Fig. 8.7 are deliberately a bit pedantic. They are met to offer core, essential, conceptual principles that apply to data and information at a pure business level. You can extract your own from this list. Refer back to Chapter 3 for more background on GAIP™. Once your initial seed list has been identified, walk through each one of these core principles to ensure coverage.

GAIP™ - Generally Accepted Information Principles	
Principle	**Description**
Content as Asset	Data and content of all types are assets with all the characteristics of any other asset. Therefore, they should be managed, secured, and accounted for as other material or financial assets.
Real Value	There is value in all data and content, based on their contribution to an organization's business/operational objectives, their intrinsic marketability, and/or their contribution to the organization's Goodwill (balance sheet) valuation.
Going Concern	Data and content are not viewed as temporary means to achieve results (or merely as a business by-product), but are critical to successful, ongoing business operations and management.
Risk	There is risk associated with data and content. This risk must be formally recognized, either as a liability or through incurring costs to manage and reduce the inherent risk.
Due Diligence	If a risk is known, it must be reported. If a risk is possible, it must be confirmed.
Quality	The relevance, meaning, accuracy, and life cycle of data and content can affect the financial status of an organization.
Audit	The accuracy of data and content is subject to periodic audit by an independent body.
Accountability	An organization must identify parties which are ultimately responsible for data and content assets.
Liability	The risks in information means there is a financial liability inherent in all data or content that is based on regulatory and ethical misuse or mismanagement.

FIG. 8.7

Generally Accepted Information Principles™

Many organizations have their own business principles, or values. Make sure you don't interfere with them. I have even seen data principles renamed to data values, to not confuse staff with business principles.

Do not forget to consider the rationale and implications and record them. This will be really valuable down the line.

Lastly, please make sure your sponsor and/or steering body approves your principles and fully supports enterprise-wide adoption.

Approach considerations
This is not a task to take lightly. We often see a list of principles lifted from a published source, plopped into a document, and then mailed out with a decree that these are the principles. These rarely succeed.

The principles need to be derived and refined formally, not casually. For a low-profile effort, either settle on a generic SINGLE principle or defer principles until a subsequent iteration (but use the first use case or iteration to demonstrate the need for principles). But when you decide to develop principles, you are starting the journey toward visibility.

Minimally invasive programs need to understand that at some point, a principle is required. There needs to be unification to the various efforts. Remember you are making something formal, which may exist in an informal manner. Business value and better data management are certainly an outcome, but long-term, you want a mind-set change, not a project success factor. So at some point, your DG team needs to formally approach principles.

The duration of this phase depends entirely on the ability of the DG team to tune the principles to their organization and get sincere and effective review from leadership. These activities can be a great opportunity to build consensus and increase the internalization of DG—or it can drag out and dissolve into a seemingly typical exercise of irrelevant meetings. If your organization starts to spend more time on "wordsmithing" principles so as not to offend anyone, you have encountered either one of two cultural issues.

1. The principles represent changes that are perceived as an admission the organization is deficient, or "bad," which the sponsor needs to assuage.
2. The participants in the process are in fear of being pointed at as instigators of bad things. The sponsor needs to make sure the team knows it is covered and someone has their back.

Do not forget to spend time with the implications and ramifications of the principles. It is only fair to be able to explain why a principle has come about (the rationale). The implications are very important, not only from a perspective of understanding, but also because implications almost always provide requirements for policies.

If you have a lot of wordsmithing associated with development of your principles, the leader of the DG deployment may need to declare the principles are good enough. They will evolve slightly anyway, so there is not much risk in starting to publicize them and begin policy development.

The general outline for a principle should contain the following elements:

1. Short description of the principle
2. Long description and full definition of the principle
3. Rationale, or a statement of why the principle is necessary
4. Implications, or statements of potential and known impact the principle will have

The most common mistake with principle development is to create policies and call them principles. If your principles are showing any of the following warning signs, you doing policy making:

- More than 10 principles: while not unheard of, when you have more than 10 principles you are starting to get very specific about what they mean.
- Using the description to declare how the principle will be enforced
- Mentioning specific business areas or functions within the principle

In general, you will start with 12–14 principles and then whittle them down to 3–4.

Ramification and benefits

Let's reinforce the importance of the review and refinement of principles. This is best illustrated by examining the results from two very different organizations (seen below in Figs. 8.8 and 8.9). My consulting practice has assisted many companies in the rollout of a DG program, but these two stand out due to the marked differences.

As the team worked with these companies to develop principles, we started at the same point. We used a "seed list" (as described above) and the GAIP™ technique. We also made sure that the principles clearly demonstrated alignment with business direction and philosophy. However, we came out with two very different sets of principles. Both are in the same industry, but the names have been altered due to nondisclosure requirements. Fig. 8.8 is a large company. Fig. 8.9 is a mid-sized company. Note the difference in tone and granularity. Both sets of principles are effective. Both sets had different ramifications to their respective organizations.

Occasionally approval of principles becomes an issue—executives feel that there are too many principles and policies. Even if it takes a while to get the principles approved, do not stop. You can continue working on other activities in this phase. The principles affect ideas, belief, and behavior. You can continue to work on the "nuts and bolts."

Sample output

The first sample is the textual version of an entire principle.

Information should be authoritative

Short description

There should exist a single, authoritative source that may be interrogated to determine any fact about any subject or object of interest.

Long description

There should exist a single, authoritative source that may be interrogated to determine any fact about any subject or object of interest. This does not preclude creating certified copies of data and information (this is understood as "managed" redundancy).

Rationale

- A verified, accurate source for an enterprise subject area data is critical to achieving comprehensive data integrity, and reduces confusion, complexity, and cost.
- While data is collected from a variety of internal and external sources, resulting in inconsistencies, these must be resolved to provide a single accurate view.
- The shift in focus from a product-centric to a customer-centric organization requires easy access to accurate and consistent data that spans functional business units.

- Common and consistent data is required to present Farfel customers with a single view of Farfel.
- Costs associated with unnecessary movement and maintenance of redundant data must be eliminated, and access latency must be reduced.

Benefits
- Reduced risk from disconnected applications projects
- Improved business alignment due to structural need to collaborate
- Reduced costs associated with data and information movement
- Reduced cost associated with departmental database proliferation
- Increased accuracy in business measures that are based on consistent data elements

Implications
- There will be a single, clearly identified, authoritative source for each managed enterprise subject area data element.
- The authoritative source and definitions will need to be easy to find and determine.
- Multiple data stores may exist within a managed environment, but one is designated as authoritative.
- Data location will be transparent to strategic business units.
- Procedural discipline (governance) is required to consistently establish this practice.
- There will be a single source of authoritative data regarding customer satisfaction and loyalty for enterprise users, dealers, vendors, field personnel, and others.
- Establish data stewardship for enterprise subject area data.
- IT and business data stewards will need to show unified support for this principle.
- Enterprise data management policies must be defined, communicated, and followed.
- The enterprise information resource must be managed with an approach that requires a centralized data management function (central authority). This function must be clearly responsible for ensuring a single, authoritative source exists for enterprise information.
- Replication and extraction must only be used as required for optimizing performance or for supporting controlled, local data updates. A governance process for replication and extraction will be required.

Below are the two aforementioned example lists of principles from two organizations in the same industry. Fig. 8.9 may have one or two extra principles, but this organization felt that policies would not hold up unless they were made part of principles.

Guiding business strategy / philosophy	Information principle name	Description
Increase shareholder value	Data and content as an asset	All BigCo Enterprise Data and Content will be managed as a corporate asset, using formal Principles to guide quality, compliance, value, and use of information.
Improve efficiency	Right person, right time, right place, right cost	Business stakeholders will get information and content delivered at the right time, location, and amount as efficiently as possible.
	Relevance	BigCo will designate federated enterprise standards and guidelines for all metrics, data structures, documents, and consent.
Business alignment and proper federation	Business alignment	Information management applications, technologies, and implementations will be aligned with business needs, and not driven by technology.
	Share and collaborate	BigCo data will collaboratively apply analytics and other uses of information to address business opportunities and challenges.
Accountability	Accountability	There will be accountability for overall integrity of enterprise data and content.
	Governance	Data that is designated as "governed" will be under the oversight of existing business areas that have appropriate authority and accountability to define and establish how information, data, and content is managed, controlled, and disseminated.
Risk management	Risk management	Management of enterprise information will reflect compliance with statutory and federal laws, policies, and regulations; such as but not limited to security, privacy, confidentiality, and data reporting.

FIG. 8.8

Data principles example

Information Principle Name	Description
Information Is an Asset	Information is an asset that will be leveraged across MidCo to improve operational efficiency, enhance competitive advantage, and accelerate decision-making.
Information should be representative	Information will represent the authentic and faithful model of MidCo's real world and its objects.
Information should be authoritative	There should exist a single, authoritative source that may be interrogated to determine any fact about an object of interest.
Information should be accurate	All available facilities, such as controls, standards, and governance, will be employed to maintain the accuracy of MidCo's information.
Information should be timely	The value of information decreases rapidly in proportion to any decrease in its timeliness and should only be retained for the duration of its useful business life.
Information should be shared	The value of information to MidCo increases in proportion to its appropriate use.
Information should be secure	Like any asset of MidCo, information should be protected from intentional or accidental corruption or destruction.
Information should be intelligible	Information at MidCo will be managed to remove risks from misinterpretation and misuse.
Information and content should be catalogued	The degree to which information is applied consistently depends on its ability to be found and shared.

FIG. 8.9

Data principles example

Tips for success

Some of the discussions around principles can become dry. Even the most enthusiastic data architect or business member of the team can start to nod off. Keep review sessions short. Also, divide the work. Have a few people write some of the principles while another group writes the others. Then have them exchange and critique.

The first pass will result in many principles that should be policy—it is not uncommon to feel you need 20 principles. But this is akin to saying there are 20 commandments or 20 amendments in the bill of rights—you dilute the power of the philosophy and belief.

Identify DG-enabled opportunities

Once you have a good idea of organization goals, needs, and how data principles will drive governance, cross reference and list what types of opportunities can be developed to implement DG in support of business needs.

Approach considerations

In this activity you will review all of the various aspects of where DG can support the organization and start to line DG up with specific initiatives or projects. From the alignment exercise you should have a list of business initiatives, or even existing projects, where DG is able to support the effort. At this point, it is time to select the most likely candidates to show value and implement some portion of DG capability.

One reason it is separated from the alignment activity is, occasionally, several active or planned initiatives are good candidates to deploy DG. Often you can combine this activity with the alignment activity. You may want to combine them, or even discuss prioritization with sponsors. Another reason is it starts a formal consideration of DG and value within leadership or sponsors.

This checklist item is short duration, regardless of approach. A low-profile effort would have already done this as part of earlier alignment, as usually a specific business need is already determined for many low-profile efforts.

Ramification and benefits

This initial, formal consideration of where DG fits into strategy is likely to be the first sign you get of "hey, you're serious, aren't you?" This is a good thing. It is certainly early enough to address. Get people into a room, review issues, and correct perceptions.

Sample output

See Fig. 8.5 for Rocky Health—as a low profile effort, they already identified the DG opportunities.

Develop business value of DG

This activity is where the DG team identifies specific financial numbers and determines what business metrics will indicate the success of DG. This is also a good place to show the cost of nongovernance or continuing to use information in a poorly managed fashion. Regardless if the team arrived at a business/data governance intersection from the first or second activity above, this activity is required to put some numbers together.

For example, if we take an example from Rocky Health, Fig. 8.5, the goal of improving patient wellness, we can ask, "If this happens, what is the anticipated impact on financial statements?" Assuming there is an intended financial benefit, then we need to look for the numbers Since we have an idea of what application of data might happen here, we can either claim the whole amount for DG *or* take partial credit based on how much of the resulting business action may or may not be directly enabled by good data. For the Rocky Health case, wellness is a key driver of healthcare. Reimbursements from government and insurers are based on results. Wellness goes up and there is an increase in revenue. Granted, we are not doing the project to deliver any data. The goal here is *not* to develop an accurate forecast of business benefits; the goal is to show that without DG the likelihood of these benefits is reduced.

Approach considerations

The lower profile DG activity may not need to produce financial benefits if the initial efforts are a proof of concept. If your DG efforts are larger or more formal, you may need to set aside some time to produce numbers and get them verified. Most often, this exercise is a matter of reloading numbers and assumptions because DG is supporting an analytics, MDM, data quality, or similar data intensive effort, and there is already some sort of business case.

Ramification and benefits

At this point, the DG sponsors and management should begin to see DG as a business program. Very often you will be challenged that the benefits are not attributable to DG. This is easily addressed with a discussion over trying to do the new effort without DG. It may add some value but will also create a new stand-alone data solution. And of course, if you are talking about DG seriously, you are trying to avoid those. At any rate, the foundation for ongoing reporting of DG value is in place.

Sample output

Fig. 8.10 shows a generic (and extreme) example where actual revenue increases were aligned with DG. Extreme, but included to show that building a solid defense for value from DG is possible.

Enterprise Goal	Information Usage as a Product-build into offerings		Information Usage as a Process-Improve cycle times, lower costs		Benefit Potential (000's)
	Levers, Actions	Objectives/Results	Levers, Actions	Objectives/Results	
Improve Customer experience	Determine level of attention (or avoid harassing) in-store shoppers Determine easier product location layout	Increase customer shopping satisfaction Increase re-visits Improved Customer Satisfaction Increased sales	Develop customer profile / score at POS touchpoints to offer promotions, affinity cards Store online information securely to avoid reentry	Increase customer shopping satisfaction Increase re-visits Improved Customer Satisfaction Increased sales Increased store/web traffic from targeted customer segments	$12,500 increment to store sales
	Data enabling the business actions				
	Advance analytics for in-store movement, product locations		Clean, secure, protected customer profile data, integrated with affinity cards and marketing		

FIG. 8.10

Business value of data governance example

Summary

Do not despair if you feel this set of activities will result in a pro forma or artificial number. After all, business benefits tend to sail right past any sort of information project and land at the feet of the business area. Even if your team is low profile, or even in stealth mode, you still have a quantitative means to monitor DG. The main benefit here is formal consideration of how DG will contribute to the business—not to data quality or other efforts—but to the business.

Strategic requirements

DG has had to evolve as data and information assets become more tightly coupled to an organization's success. Where a DG program might have been more focused on supporting a data initiative 10 years ago, monetization of data and insertion of digital strategies into corporate strategies has required DG programs to become more strategically engaged. This set of activities addresses taking a strategic look at DG. These activities are part of the overall engagement checklist. You may or may not need these activities. You may need some, and not others. For sure though, every DG program that is restarting or getting started needs to run the checklist and see if any of these activities are necessary.

Review existing business cases

When DG gets started in connection with a particular initiative, such as advanced analytics, artificial intelligence, MDM, etc., it can become more of a part of the initiative vs a strategic capability. This is something I see in programs that need to restart. For example, DG was initiated to improve data quality for analytics. That was done, everyone looks to the next big thing, and DG loses its status as a business capability. It then becomes hard to sustain.

In order to avoid getting set aside as a completed task, consider reviewing all existing business cases and initiatives. There are most likely projects on the drawing board that require some sort of data or information. And more often than not, no one has considered a role for DG. You will likely find similar efforts that can benefit from the very use case or low-profile effort you are contemplating. In addition, you may find a significant effort being planned where management needs to consider DG as a required component.

Approach considerations

There isn't much to consider—if you are restarting a DG effort you need to look around at other activity to make sure you are getting the full picture. Low profile efforts may want to look around or focus on the known task at hand for the time being.

Ramification and benefits

Every single time I have been with a client and we have performed this activity we have found additional initiatives where DG needs to be applied. Even if we need to focus on a specific use case, are in stealth mode, or doing a large, invasive, regulatory-drive program, we have had to create a list of other initiatives where DG needs to happen. More important, the nature of the DG work we might currently be undertaking applies to other initiatives. Worst-case scenarios can happen where the current DG work will actually hinder or affect another effort. No matter what the nature of conclusions of this exercise, additional communications with leadership may be in order.

Sample output

Nothing to see here—anything that comes out of this needs to be added to alignment or business case deliverables.

Determine base line policy requirements

This activity is the step where you can identify new strategic policies. There are obvious policies, usually extracted from the initial principles. There will be another area where process and policy are examined in more detail. This is an opportunity to identify policies that are obvious and have a strategic impact. An example would be a principle declaring that all significant data will have a defined and certified source. Obviously, there are policy implications that can be identified as strategic. Often in MDM the process changes required to implement MDM are significant. As part of the Strategy work it is good to understand these.

Approach considerations

The actual activity occurs along parallel efforts. Fig. 8.11 shows how the principle provides input and inspiration for other components of DG. The information principles need to be evaluated for implied policies.

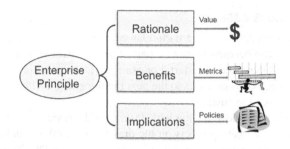

FIG. 8.11

A taxonomy for data principles

Ramification and benefits

The primary benefit is the ability to view potential DG activity as strategic. Of course, the downside could also be that the business gets to see what is required to operate the DG program!

Sample output

Here is an example of initial policies based on the Rocky Health narrative (Fig. 8.12).

Principle	Initial Strategic Policy (Simple, Foundational)
We must be accountable for managing data as an asset that improves patient outcomes.	Projects and report development will include data management in project plans. Data usefulness and accuracy will be equally or more important than delivering reports or completing IT projects.
Data usage and sources will be transparent with a defined source.	Rocky Health will define common standards for data definition, sourcing, and access.

FIG. 8.12

Principles and policies example

Tips for success

There may be push back along the lines of "we are not ready for this." You have moved from abstraction to reality and must be prepared to address the concerns. A few valid responses are:

- It looks like a lot because it is new, but you already do similar things.
- New policies are part of the inevitable changes of DG. So we need to ensure we address the change issues.
- I understand your concern—perhaps we need to adjust our approach, but it is already as low profile as can be—so do we need to defer solving these data issues?

Identify obvious capabilities

Remember capabilities are an excellent means to shape and convey the "What" aspect of DG.

During the Engagement activity, we used an initial identification of capabilities to provide reinforcement for a vision statement. If you did that, you may have already addressed what you need. This activity can use those to confirm them as genuine requirements. You may have performed the Engagement activities and not addressed capabilities for some reason, and while reviewing the checklist, now is a good time to present obvious capabilities.

Even a noninvasive effort needs to, at minimum, start to point out what DG capabilities will be stood up as part of your low profile DG activity.

Approach considerations

The theme here is to use the checklist approach, then identify required capabilities as soon as they start to appear. If you have not identified obvious capabilities yet, have the alignment activity and the business case activity to infer required capabilities.

Ramification and benefits

Don't become consumed if you know obvious capabilities already or are restarting a program and already have DG capabilities that require reinforcement. Just remind your audience of these items. We are not doing capability modeling yet. The benefit is the same as if you did this in Vision; you get an obvious and easy-to-see view into DG. You may need additional details to associate the capabilities with the use cases that might arise in this work area. Remember the focus in this area of work is strategic; capabilities are the best means to describe and manifest long-term elements of your DG program.

Sample output

Refer again to Figs. 8.4 and 8.5. Both show capabilities aligned with DG along with benefits. This type of alignment is what any size effort should try and accomplish.

Identify use cases to show value (as required)

After many years of DG work, it is very apparent that most organizations cannot move ahead with DG as some sort of foundational project. The typical scenario has been "acquire a glossary tool, load it up, then start to use it under the auspices of DG."

There are some problems with this approach. Even if you are blessed with enlightened leaders, it is not agile. It is linear and sets up very rigid dependencies before any value can be seen. You have no value to show any resisters and detractors. You are therefore exposed to changes in business conditions. Simply, your DG efforts will be the first thing canceled with a down turn of fortunes. It appears too much like overhead.

Enter the use case. During my time at First San Francisco Partners there was a lot of work done with organizations that simply had no other choice than to show value quickly or lose what minimal credibility DG had developed. While early, value-added iterations were always part of a roadmap, we developed specific techniques to draw out these value-added iterations, that is, use cases.

Helpful hint

Use case replaces "low-hanging fruit." I hate that phrase. It has become a cliché, and a go-to catch-all when the sponsor feels the heat. The big boss leans in, and suddenly your well-thought DG effort needs to address "low-hanging fruit." Your budget to do DG is now being allocated to do other things.

That kind of tactical effort is NOT THE RESPONSIBILITY of a DG or even data management team. Low-hanging fruit, or tactical things, are the responsibility of the Big Boss and need to be scheduled outside of transformational efforts such as DG.

If your culture is ripe with low-hanging fruit (pun intended) then get ahead by making sure your use cases proactively address any low-hanging opportunities. But keep them as part of your program and get the full credit.

If, after reviewing the checklist of possible activities, you may feel some use cases are necessary. Executing this activity should provide a handful of options that will add visible value (i.e., solve a problem); in addition, you will need to deploy some sort of DG capability to ensure the solution is completed and is sustainable.

The use case development merges alignment, business strategy, required capability thought, and perhaps even other planned work. While the examples show there is some mechanical aspect to doing this, by and large, it is an analytical activity.

Approach considerations

At this point, larger efforts will have a collection of alignment and strategic artifacts. Review alignment and strategic documents for specific opportunities where data must be used. This may have been down already to look for capabilities and obvious use cases. If so, you can skip this, or use a confirmation activity. Fig. 8.13 shows the synthesis of various inputs.

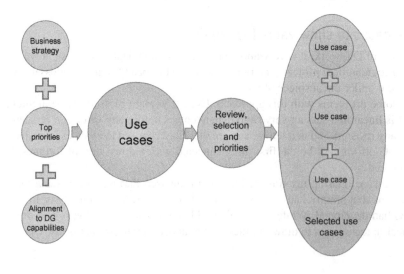

FIG. 8.13

Use case process

It is important, that whether use cases start here, or earlier, or even later, that you make sure that completion of the use case results in visible benefit. Verify that a real contribution to value will result. Again—loading a glossary will not appear as value-add use case.

Finding some use cases will also point to DG and management capabilities that will be required to support the use case. Again, you may have some already, and see more. Or this may be the first pass. It all depends on the size of your efforts, or even if you are making a second attempt at DG. Identify required DG and DM capabilities.

The difference in dealing with use cases now vs Engagement is level of detail. There is now the need to get specific about what the use case is about. Fig. 8.5 in the Rocky Health case study shows a common template. You need to be able to clearly show both benefit and type of activity (that is most likely NEW) that will accompany the use case. But the details need to be sufficient so others can review and approve use cases.

Use cases will need to be presented to leadership at this point. There may be several types of "asks": prioritization of many possible use cases, selection of one, intervening with a PMO or development group that may be working in that use case's region, or even getting funding if resources are required.

Ramification and benefits
Reviewing use cases as part of strategy accomplishes several things:

1. Leadership is more or less forced to consider "projects" that will add value, but also carry along implementation of new DG (and usually data management) capabilities.
2. The DG team, whether it be one or several persons, will need to speak in business terms, and eloquently indicate the subtle difference between a DG use cases and "run-of-the-mill" projects.
3. Raises awareness of DG and increases engagement. Or, it raises issues that can be addressed sooner than later.

Sample output
Fig. 8.14 shows a generic retail derivation of use cases. It is connected to the prior example where the actual target increase in revenue was estimated. Goals are aligned with uses of information. Then, in lieu of access to well-documented business strategy, some general uses of data for value were applied. The result was nine possible use cases. This is a redacted example of a real use case exercise. There was no available business strategy, but notice how some very effective use cases were identified by applying industry trends.

Information use as a Product - data built into offereings					
Related Goal	**Data use, or data lever**	**Possible process or scenarios of information use (use case)**				
Improve effectiveness of website	Improve on-line ordering experience with access to order history	Provide means to enter preferences and profile purchase information (cr card, etc.)	Provide Customers the ability to access complete Order History to reorder or reuse part of an order	Store history and customer contact data	Enable Repeat Purchasing Based on Order History Lookup	Provide customer with similar item offers

Information use as a Process improvement - improve cycle times, lower costs					
Related Goal	**Data use, or data lever**	**Possible process or scenarios of information use (use case)**				
Improve Customer experience	Supply customer profile/score at POS touchpoints, offer promotions, affinity cards, repeat purchase ease	Store history and customer contact data	Provide ability to access complete order history (in store web access)	Evaluate histories to design offers and promotions	Offer promotions or affinity items at POS	

FIG. 8.14

Use case derivation

Strategy case study—Rocky Health

While the engagement activity's main role is to ensure the organization will support DG, Tom needs to look ahead and finalize an approach. This means looking at the Strategy and subsequent checklists to determine how he will show value. Remember, he is an army of one. Everything he asks of the organization needs to have some visible benefit. Tom understands even a low-profile effort needs some sort of strategy, design, and rollout plan. Tom and the consultant ensured that his peers at Rocky Health were engaged. But the engagement was partly due to the recent successes. Expanding DG meant there had to be a plan for keeping the organization engaged in DG. The modest DG effort (Tom, the part-time resource, and the mentoring consultant) had to look ahead at the other checklists to leverage an approach.

This is how he determined his sprints. Two sprints were identified with multiple purpose. First, to continue to show value of DM and DG incrementally. But Tom remembered that management approved "expansion" of DG. The sprints needed to cover that as well.

Refer to Fig. 8.5. The use cases chosen will then help determine what is done in terms of the next areas of work—Architecture and Design.

Strategy case study—Rocky regional electric coop

The planning of the approach is longer than expected. Diana had surmised maybe a week. But it took 3 weeks. First there were the three situations that were front of mind. They had been labeled by the team as:

1. Enterprise reporting and business intelligence (BI)
2. Engineering and asset management
3. Customer service hook up

Diana explained to her team during engagement there was a lot more to enterprise DG than these three issues. Addressing just these three areas under a label of EIM would only create a modest capability to manage data, and essentially leave RREC with yet another set of stand-alone solutions. In addition, BI and reporting is not really a business use case—where is the companion-specific business goal?

Their planning also had to consider the bad feelings from prior efforts. Showing value early was going to be critical. Just like Rocky Health, there has to be a balance of short-term results with long-term sustainability.

Diana had to use the alignment and capability activities to develop a strategy that not only showed value, but she had to tell the organization that DG is a new enterprise level business capability, especially with the use of industry standard models. Having a business area create data standards will not be a bad path, but the standards need to be properly installed, used, and administered. She does not want to frighten off support, so this all needs to be "baked in" to her solutions to the problems the team needs to address.

The alignment exercise (Figs. 8.4 and 8.15) was more important to provide value and support. After all the use cases were more or less self-defining. They are the three topics: Enterprise Reporting and BI, Engineering and Asset Management, and Customer Service Hook-up. Diana had to make sure these were business focused efforts. The three initiatives are now framed in their value to the business, and the capabilities they need from DG.

Goal	Documented Objectives	Initiatives	Use case candidate	Business Data Mgmt Needs	Possible DG Capabilities
Reduce Costs	Reduce tool and material redundancy 25%	Minimizing the tools & equipment needed to operate our business	Equipment and Engineering accuracy	Tool and equipment data management	Data lineage, data quality
Maximize return on assets (plant, people, processes)	Reduce new transformer installation cycles by one week	Attain more efficiency in asset set up and management		Tool and equipment data management	Data standards
Increase Value to Customer	Reduce new connection time to one	Improve responsiveness (cycle time, on time, information, etc.)	Customer service installation	New service appointments, scehduling and asset availability data	Item, Inventory accuracy, data quality
	Increase customer satisfaction on new service to 90%	Deliver service to expectation levels			
			Business Intelligence and Reporting	This is a data management capability - very often I see this as a use case, but it is not a good candidate. Add the BI and Reporting capability to other business use cases	

FIG. 8.15

Rocky Energy use cases

Summary

Efficiency is doing things right; effectiveness is doing the right things.
Peter Drucker

The strategy for tackling DG is a function of alignment, policy, and business value. You can and should have a strategy for DG, even if you are minimally invasive.

Alignment gets you to business value and required capabilities and use cases. Do not despair if you feel this set of activities results in pro-forma, or deduced statement of value. After all, business benefits tend to sail right past any sort of information project and land at the feet of the business area. Even if your team used a guerilla approach, you still have a quantitative means to monitor DG. The main benefit here is formal consideration of how DG will contribute to the business, not data quality or other effort, but to the business.

Relevant principles of information management are defined. These then frame the statement of implications and rationale. These then frame policy development. We are not done yet; we still need to identify "who" and "where" the DG happens. But we have a firm grasp on the "what." From the principles to the capabilities and use cases, you need to use these activities to develop a clear statement of what DG needs to address to add value to the organization.

Some readers may be thinking that Diana's approach is too aggressive because it seems she is taking on all of the enterprise initiatives at once. Remember this topic is STRATEGY. What needs to be done to support the organization? What are the obvious capabilities and requirements? Regardless if there are large enterprise-wide drivers, or just simple projects (use cases) like in Rocky Health you will need to be iterative with your implementation.

We have not said what kind of implementation is to be done. We still need to design the DG operating models. We cannot plan implementation until we know how DG is supposed to work. The next area of work, Architecture and Design, will supply the design of the operating elements. Regardless of a short design effort, or long one (due to complex situations), we need to determine how the DG program should operate. Then we can move into planning implementation.

Essential questions

1. Strategy does not necessarily need to spend any tie on aligning to business needs if you are doing a simple, low profile implementation of DG. True or false? Defend your answer.
2. Describe how your DG team would approach alignment if your business strategy was not readily available.
3. Why do you need to specify use cases, or small, incremental units of work, to implement DG?

Architecture and design

Chapter Outline

The ability to convert ideas to things is the secret to outward success.
Henry Ward Beecher

Overview

This chapter covers the activities related to designing the data governance (DG) program; that is, how things need to be done. Also, there are tasks offered to appoint the "who" along with roles and responsibilities. In short, this work area, which covers a lot of activity, contains what need to be done to move from strategy to a defined DG program, suitable for your situation, and ready to be applied to a roadmap or roll-out plan.

Data Governance. https://doi.org/10.1016/B978-0-12-815831-9.00009-6

Some very significant program artifacts are developed that will remain with, and be used by, the DG program for as long as it is in operation.

Remember the analogy of the DG "V" when determining what to do in this work area. Your team will be addressing information or data management (DM) capabilities and functions, *in addition* to the DG capabilities and functions. What is really happening in this phase is the *formalization of the operating model for DG in support of DM*. This builds out the DG capability into something that can be implemented. See Fig. 9.1.

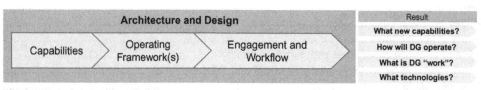

FIG. 9.1

Architecture and design work area

An obvious question might be, "Why should we build the DG program by identifying data or information management activity?" The answer was stated earlier—you have to govern something, and information or data management is the main subject. You need to specify what is governed. You also need to be very clear about what the DG program has to do and who is accountable. The eventual makeup of the DG operating framework depends on a well-defined result from these activities. If DG is to be a "net zero" or minimal increase in cost, where does the oversight come from? If the V is focused on separation of duties, don't we need more people to handle oversight while others are working on information projects? In reality, very few organizations can afford to add a 100%-dedicated DG staff. (Those that can are unwilling.)

The answer is found in how you assign the various duties and desired new behaviors. For example, if a business leader is sponsoring a master data management (MDM) project, then obviously that person cannot be responsible for the DG of that project. If a chief analytics officer is overseeing an advanced analytics area, then oversight needs to come from somewhere else. Therefore, another business leader is given the duty to provide oversight. This is not an organization—it is an operating model, or description of new behaviors.

Early in the days of doing "stuff" called DG, we did not pay much attention to how the organization would operate after all the new data, tools, and neat things were turned on. After all, how hard could it be? Appoint a few stewards, give them power to enforce standards, and away we go. It was not long before we realized there had to be formal engineering to show people what needed to be done. In the ensuing years, the operating framework has developed into an artifact almost as crucial as a roadmap. I could argue it is more important, as understanding how it works is more important than understanding when it will start to work.

Also, note that the term "operating framework" is used instead of "organization chart." Avoid use of the term "organization" in the context of a noun. You are never building an organization in the context of a new department. The goal is to eventually blend in with ordinary day-to-day behavior; you will rarely develop a large separate DG organization.

> One of the most interesting phenomena I have seen is the political struggle to get control of the DG "department." The CIO says the new department needs to be in IT, while the risk officer argues it needs to be in Compliance. In truth, the best DG operation is accomplished with cross functional operating models. However, don't fool yourself that this scenario means you are dealing with politics. Politics is a symptom. Power grabbing is a symptom of insecurity. The real issue is not politics, it is ignorance of what DG really means. It is an indicator of deeper issues of understanding. Politics will always find you; you cannot be proactive about it. But you can be proactive about change.
>
> For organizations with large efforts, or high visibility, the organization change work should be well underway during these activities in Architecture and Design. I will address this more during the overview of sustaining the DG program.

This area of activity also entails identifying the stewardship/ownership/custodian population. We nail down what will be done where, and who will do it. The emphasis in these checklist items is to identify participants with a clear understanding of how they will be participating. In some cases, *there may or may not be formal stewards* over the long term. But most importantly, if you have defined stewards before you have described an operating framework, you have given responsibility without:

- "Air cover" of a defined process
- Formal means to monitor activity
- Training
- Organizational acceptance of one definition of "steward"

This chapter divides the activities into three topics—Capabilities, Operating Frameworks, and Engagement and Workflow. This chapter also addresses organizational change management as a capability and part of DG operations.

- Capabilities focuses on finalizing what DG will need to do. I have added technology to this area. DG technology enables DG capabilities. It's not a separate topic. I did this deliberately to make sure the two areas are considered at the same time.
- Operating frameworks covers the How, and the organization (verb, not a noun) of responsibilities within the DG program.
- Engagement and Workflow will cover the activities required when you need to define how DG works with other areas, and/or provide detailed roles and responsibilities (e.g., stewards) as well as reinforcing the change process, socializing these new roles and engagement processes.

Remember, your particular initiative may not require all of the activities in this chapter. This is a checklist. You need to select and apply what is relevant for your situation. The variety of activities can be seen in Fig. 9.2.

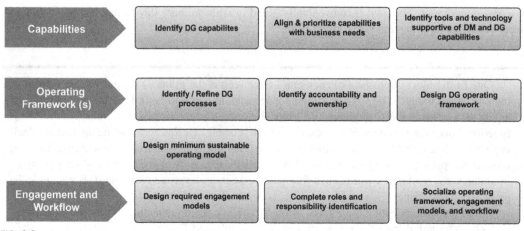

FIG. 9.2

Architecture and design activities

Capabilities

Revolutions always come around again. That's why they're called revolutions.
Terry Pratchett

This set of activities is used to finalize the WHAT and initiate the HOW. As you know, we have discussed capabilities twice already—where identifying capabilities offer insight into engagement and strategy. However, this area is where you need to nail down what will really happen.

Once identified the capabilities need to be aligned with business needs. This is for timing. When the business needs the capability, then roll it out. Not before.

Lastly, technology capabilities are addressed, alongside all other capabilities. This is quite deliberate. Like other capabilities, align the capability with the need, then determine timing. This avoids everyone running out and buying stuff before it can be sustainably used (Fig. 9.3).

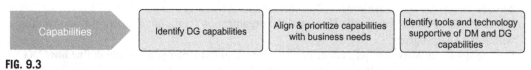

FIG. 9.3

Capabilities

Identify DG capabilities

Earlier (in Engagement and Strategy) we mentioned listing capabilities, usually if they were obvious or from a predetermined list. Now that we need to, possibly, go into a lot more detail, you need to have a process for determining and even creating the capabilities required for your scenario, whether is it a large or small DG effort.

Remember, capabilities are a lot closer to business and enterprise architect thinking. Capabilities are the language of enterprise architects, and I have noticed more relevance in getting DG engaged with enterprise architects, as well as the usual constituents.

In the first edition we called out FUNCTIONS, to indicate the WHAT, and the function lists are still in the appendix. And you still need them to design detailed processes. However, Capabilities fits better for planning and design of the program. Functions and processes are deliverables from implementation.

The basic approach to standing up DG capabilities is handled like standing up any other new business capability or upgrading current capabilities (to include better data behaviors). Capabilities are also easier to communicate to leadership. Processes and functions are too detailed. Let's review the thought process below:

Capabilities = the WHAT; what does DG need to do to enable the organization to accomplish its goals? These can be decomposed into:

Functions = Logical grouping of processes, usually in a business area. For some DG efforts, this is enough. Otherwise, if you need more details, you need to define…

Processes = Processes take some input, alter it, and produce an output. In this case, DG things get done. For an operating model, processes can also be placed into a …

Workflow = portray processes across functional areas with an emphasis on the handoffs, cooperation and collaboration via sharing work products and artifacts.

Fig. 9.4 presents a DG capability model.

Approach considerations

Capabilities need to be neutral from an organization chart view. At this point please don't worry about who will do anything. Bear in mind, no matter what type of approach you are taking, that your mindset is to elevate data behaviors across the entire organization.

Start with the standard list of capabilities (see Fig. 9.4). Also, review the DM capability list (see Fig. 9.5) and make sure you are not thinking of a DM activity without corresponding DG support.

You may find that you feel you want every single capability. First, try and be reasonable—you probably do not need every possible DG capability, e.g., few organizations apply DG to unstructured content, even though they should. Even if you do add too much, the next activity we present will help you adjust—so see the alignment activities below.

Ramification and benefits

Capability-driven design of DG has a number of advantages. We touched on these earlier, but here is a bit more in-depth treatment.[1]

1. It's a top-down, whole-of-organization approach. It breaks through departmental silos by shifting from a functional view to a capability view.
2. It focuses directly on what an organization needs to do to execute its strategy.
3. It provides a map of the organization's overall capabilities to ensure nothing is missed.

---

[1]ChuenSeet, 2018, What is Capability-Based Planning?, https://www.jibility.com/what-is-capability-based-planning/.

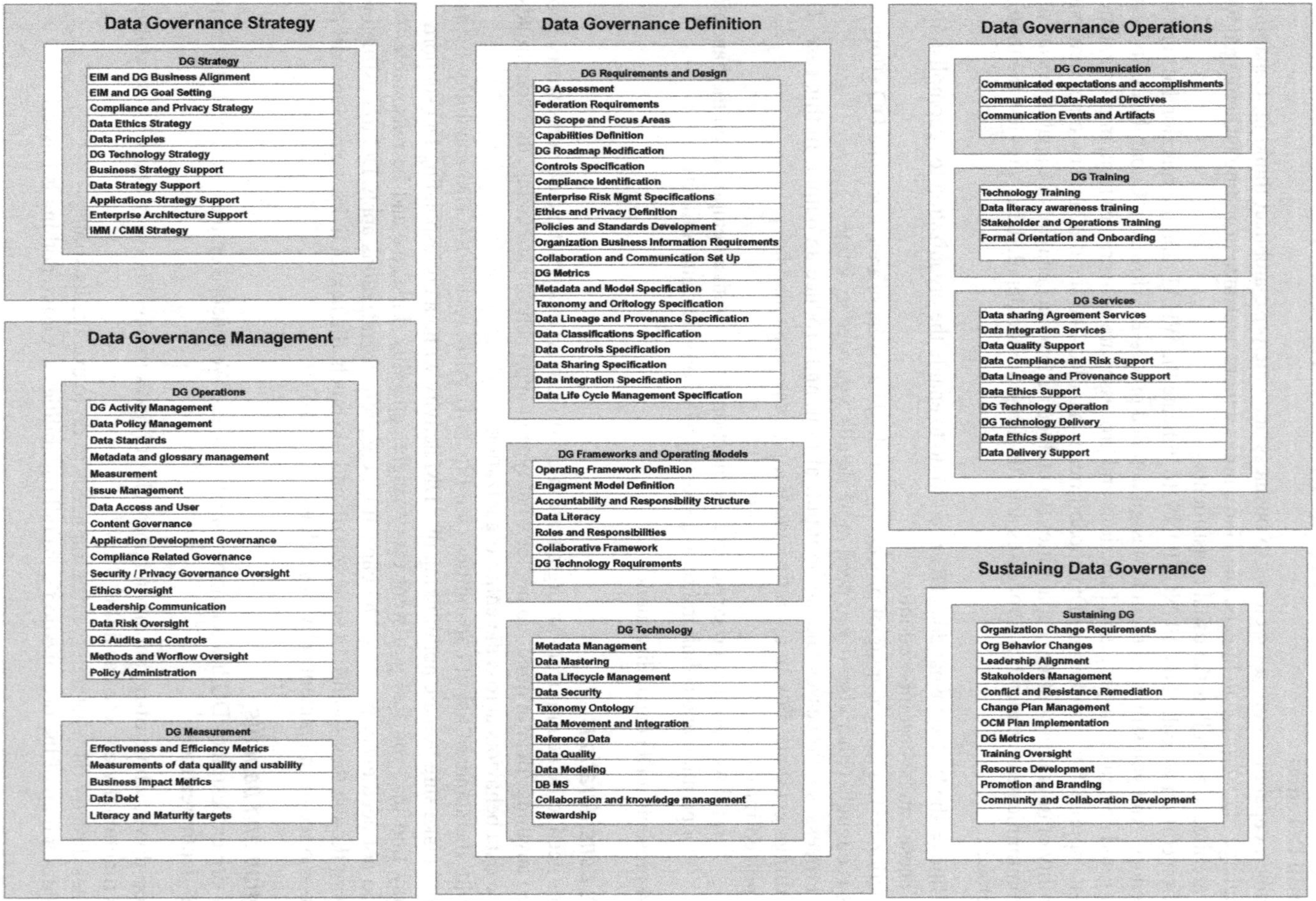

Data Governance Strategy

DG Strategy
- EIM and DG Business Alignment
- EIM and DG Goal Setting
- Compliance and Privacy Strategy
- Data Ethics Strategy
- Data Principles
- DG Technology Strategy
- Business Strategy Support
- Data Strategy Support
- Applications Strategy Support
- Enterprise Architecture Support
- IMM / CMM Strategy

Data Governance Definition

DG Requirements and Design
- DG Assessment
- Federation Requirements
- DG Scope and Focus Areas
- Capabilities Definition
- DG Roadmap Modification
- Controls Specification
- Compliance Identification
- Enterprise Risk Mgmt Specifications
- Ethics and Privacy Definition
- Policies and Standards Development
- Organization Business Information Requirements
- Collaboration and Communication Set Up
- DG Metrics
- Metadata and Model Specification
- Taxonomy and Ontology Specification
- Data Lineage and Provenance Specification
- Data Classifications Specification
- Data Controls Specification
- Data Sharing Specification
- Data Integration Specification
- Data Life Cycle Management Specification

DG Frameworks and Operating Models
- Operating Framework Definition
- Engagement Model Definition
- Accountability and Responsibility Structure
- Data Literacy
- Roles and Responsibilities
- Collaborative Framework
- DG Technology Requirements

DG Technology
- Metadata Management
- Data Mastering
- Data Lifecycle Management
- Data Security
- Taxonomy Ontology
- Data Movement and Integration
- Reference Data
- Data Quality
- Data Modeling
- DB MS
- Collaboration and knowledge management
- Stewardship

Data Governance Operations

DG Communication
- Communicated expectations and accomplishments
- Communicated Data-Related Directives
- Communication Events and Artifacts

DG Training
- Technology Training
- Data literacy awareness training
- Stakeholder and Operations Training
- Formal Orientation and Onboarding

DG Services
- Data sharing Agreement Services
- Data Integration Services
- Data Quality Support
- Data Compliance and Risk Support
- Data Lineage and Provenance Support
- Data Ethics Support
- DG Technology Operation
- DG Technology Delivery
- Data Ethics Support
- Data Delivery Support

Data Governance Management

DG Operations
- DG Activity Management
- Data Policy Management
- Data Standards
- Metadata and glossary management
- Measurement
- Issue Management
- Data Access and User
- Content Governance
- Application Development Governance
- Compliance Related Governance
- Security / Privacy Governance Oversight
- Ethics Oversight
- Leadership Communication
- Data Risk Oversight
- DG Audits and Controls
- Methods and Worflow Oversight
- Policy Administration

DG Measurement
- Effectiveness and Efficiency Metrics
- Measurements of data quality and usability
- Business Impact Metrics
- Data Debt
- Literacy and Maturity targets

Sustaining Data Governance

Sustaining DG
- Organization Change Requirements
- Org Behavior Changes
- Leadership Alignment
- Stakeholders Management
- Conflict and Resistance Remediation
- Change Plan Management
- OCM Plan Implementation
- DG Metrics
- Training Oversight
- Resource Development
- Promotion and Branding
- Community and Collaboration Development

FIG. 9.4

Data governance capability model

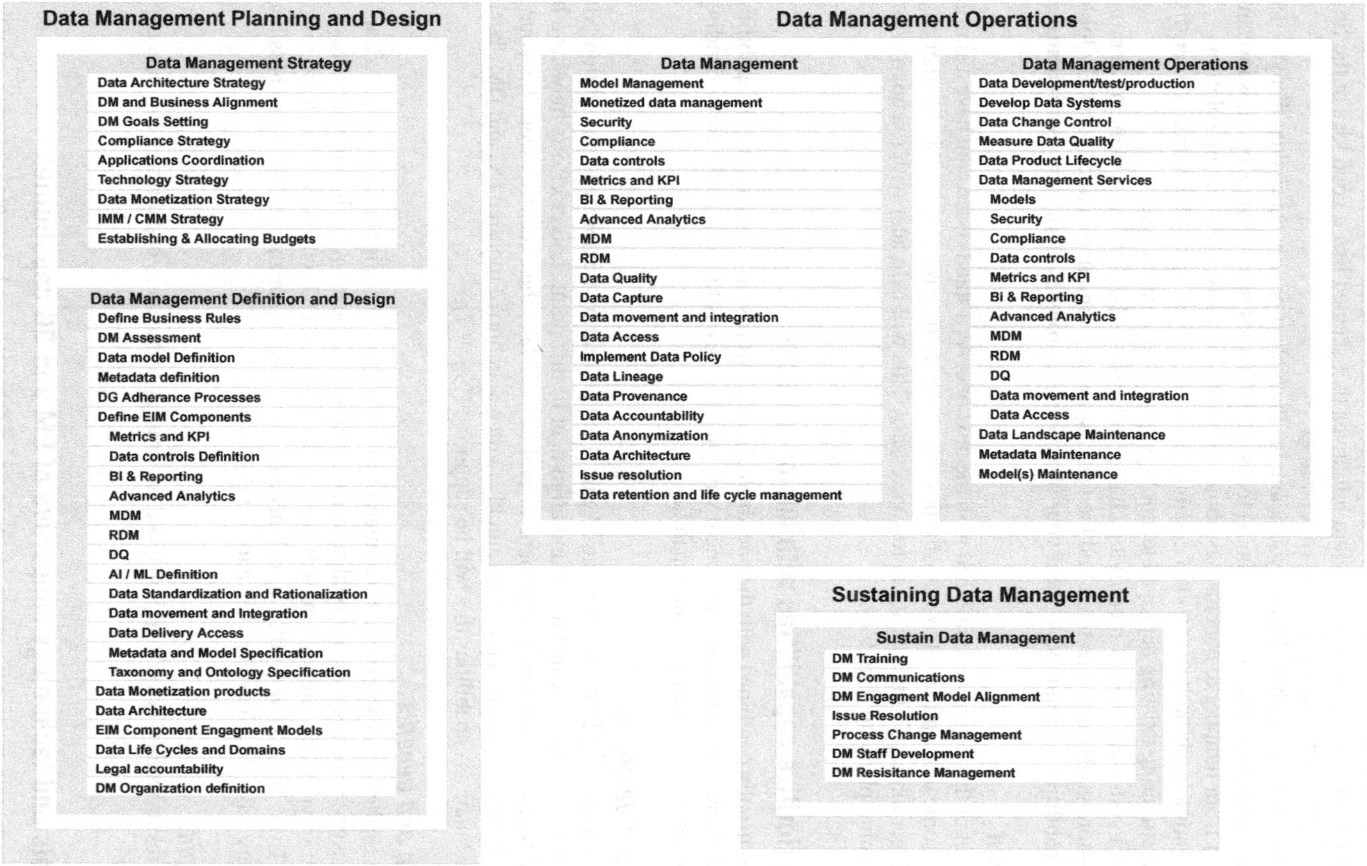

Data Management Planning and Design

Data Management Strategy
- Data Architecture Strategy
- DM and Business Alignment
- DM Goals Setting
- Compliance Strategy
- Applications Coordination
- Technology Strategy
- Data Monetization Strategy
- IMM / CMM Strategy
- Establishing & Allocating Budgets

Data Management Definition and Design
- Define Business Rules
- DM Assessment
- Data model Definition
- Metadata definition
- DG Adherence Processes
- Define EIM Components
- Metrics and KPI
- Data controls Definition
- BI & Reporting
- Advanced Analytics
- MDM
- RDM
- DQ
- AI / ML Definition
- Data Standardization and Rationalization
- Data movement and Integration
- Data Delivery Access
- Metadata and Model Specification
- Taxonomy and Ontology Specification
- Data Monetization products
- Data Architecture
- EIM Component Engagment Models
- Data Life Cycles and Domains
- Legal accountability
- DM Organization definition

Data Management Operations

Data Management
- Model Management
- Monetized data management
- Security
- Compliance
- Data controls
- Metrics and KPI
- BI & Reporting
- Advanced Analytics
- MDM
- RDM
- Data Quality
- Data Capture
- Data movement and integration
- Data Access
- Implement Data Policy
- Data Lineage
- Data Provenance
- Data Accountability
- Data Anonymization
- Data Architecture
- Issue resolution
- Data retention and life cycle management

Data Management Operations
- Data Development/test/production
- Develop Data Systems
- Data Change Control
- Measure Data Quality
- Data Product Lifecycle
- Data Management Services
- Models
- Security
- Compliance
- Data controls
- Metrics and KPI
- BI & Reporting
- Advanced Analytics
- MDM
- RDM
- DQ
- Data movement and integration
- Data Access
- Data Landscape Maintenance
- Metadata Maintenance
- Model(s) Maintenance

Sustaining Data Management

Sustain Data Management
- DM Training
- DM Communications
- DM Engagment Model Alignment
- Issue Resolution
- Process Change Management
- DM Staff Development
- DM Ressitance Management

FIG. 9.5

Data management capabilities

4. It directly links initiatives and projects back to capability changes and, in turn, back to the organization's objectives. No more random initiatives that seemed like a good idea at the time, but in hindsight don't actually align to your strategy.
5. It cuts the wheat from the chaff. It helps you determine the highest priority capabilities that you need to develop, and related initiatives that you should focus on. It clarifies and optimizes business investment.
6. It stops you from jumping to conclusions about solutions too early. By delaying solution definition and doing it in the context of capabilities, it opens you up to alternatives rather than simply incrementing existing deployed technology, processes, and people.
7. It provides a systematic way of identifying change initiatives. Many business planning approaches define mission, goals, and objectives, and then start spawning initiatives and projects. By looking at what capabilities are required to meet your objectives, it provides clarity for your initiatives.

Sample output

Often the output of this activity is combined with the next one. As a stand-alone, you have a simple list. In many cases some descriptions of the capabilities is in order, as they may seem new. Another angle on this is to mark certain capabilities as absolutely necessary. This will help later with roadmap prioritization and sustaining planning (Fig. 9.6).

Align and prioritize capabilities with business needs

This activity is usually combined with the one above. In other words, while the required capabilities are being detailed, there is an associated alignment with business needs (as well as the corresponding DM aspects). This confirms relevance of the capability.

Approach considerations

Why separate this activity from the prior one? Again, I am presenting flexibility in this edition. Smaller efforts may have delayed any capabilities and alignment to this point. Larger efforts may have had to iterate several times and confirmed alignment several times. Make the checklists work for your situation. Most of the time this task can be done quickly, with a matchup of required DG and DM "business" capabilities to business capabilities that will be supported (Fig. 9.7).

Ramification and benefits

As you move further on to an eventual roadmap or prioritization, the alignment to business capability will make it easy to prioritizes and rationalize what could seem to be an overwhelming amount of work. At some point in your DG journey you will need to declare how DG will interact with the rest of the organization. Even if you start small, you will need to exhibit an organization-level awareness one day, and this activity will add a lot of credence to that discussion.

Sample output

Below is the result when Rocky Health mapped its strategy to business capabilities and data capabilities (Fig. 9.8).

Identify tools and technology supportive of DM and DG capabilities

This section concerns itself with some of the more mechanical and discrete elements of your DG architecture. If you determined at some point technology is desired or required, this section provides the checklist items for selection and fitting new technology into your data architecture.

The Data Strategy Capabilities Model is copyright 2019 John Ladley and Sonrai Solutions. Reuse, reproduction, private or commercial use, is prohibited without explicit permission from John Ladley

Data Governance Capabilities	Required	Capability Checklist	These Capabilities are bare minimum.
PLAN		Data Governance Strategy	
	x	EIM and DG Business Alignment	
		EIM and DG Goal Setting	
		Compliance and Privacy Strategy	
		Data Ethics Strategy	
	x	Data Principles	
		DG Technology Strategy	
	x	Business Strategy Support	x
	x	Data Strategy Support	x
		Application Strategy Support	
		Enterprise Architecture Support	
		IMM / CMM Strategy	
DEFINE		Data Governance Definition	
	x	DG Assessment	
		Federation Requirements	
		DG Scope and Focus Areas	
	x	Capabilities Definition	
		DG Roadmap Modification	
	x	Controls Specification	x
		Compliance Identification	
		Enterprise Risk Mgmt Specifications	
		Ethics and Privacy Definition	
	x	Policies and Standards Development	x
		Organization Business Information Requirements	
		Collaboration and Communication Set Up	
		DG Metrics	
		Metadata and Model Specification	
		Taxonomy and Ontology Specification	
	x	Data Lineage and Provenance Specification	
	x	Data Classifications Specification	
	x	Data Controls Specification	x
		Data Sharing Specification	
	x	Data Integration Specification	x
		Data Life Cycle Management Specification	
		Data Governance Frameworks	
	x	Operating Framework Definition	
		Engagement Model Definition	
		Accountability and Responsibility Structure	
	x	Data Literacy	x
	x	Roles and Responsibilities	x
		Collaborative Framework	
	x	DG Technology Requirements	

FIG. 9.6

Sample capability list

FIG. 9.7

Aligning capabilities

Strategy Mapping		
Business Capability	Data Governance Capability	Supported Data Management
Exceed patient expectations		
Care experience	Metrics management, data life cycle, Data sourcing	BI (real time), data quality
Care quality	Metrics management, data life cycle, Data sourcing	Operational, real time update, scheduling performance analysis and management
Reputation management	Metrics management, Data sourcing	Analytics
Efficient access	Data standards, data quality, Data life cycle	Patient MDM and DQ, Periodic reporting, use for analysis, operational access
Provide best practice at best cost		
Move to value based healthcare	Metrics management, data life cycle, Data sourcing	Analytics, ad hoc analysis , Standard reports
Define result based healthcare	Metrics management, data life cycle, Data sourcing	
Localize offerings	Data standards, data quality	Reference data, Master data
Select appropriate partners	Metrics management, data life cycle, Data sourcing	CMS Metrics, Periodic reporting of performance metrics
Ensure financial stability		
Manage operating margin	Metrics management, Data sourcing	Periodic reporting of compliance and periodic metrics
Cost Management	Metrics management, Data sourcing	Periodic reporting of compliance and periodic metrics
Regulatory Compliance	Data complianance, Data provenance	Periodic reporting of compliance and periodic metrics

FIG. 9.8

Rocky Health capability alignment

Granted, most of DG is policy, change management, and workflow. But if you understand DG as a program, and implement it as such, sooner or later the prospect of getting tools of some sort will present itself.

As stated previously, this does not mean buying tools and technology is a high priority. The human and workflow elements are way more important. Nearly every DG effort I work with acquires tools too soon. Buying a glossary product first before even having an operating framework that describes how to effectively USE the glossary, is plain silly.[2]

But eventually you are going to find yourself up to your hips in digital files, documents, and other artifacts representing the multitude of "stuff" being governed. For example:

1. Data provenance is always changing, and in many cases, those changes must be tracked.
2. Just the workflow around agreeing on a data element or metric definition can be complicated.
3. You also need to administer the workflow. Workflow is not self-regulating and requires tuning and adjustments.
4. It will be necessary to administer the governance program and use some sort of automation.

There's a tendency to think that DG tools are just repurposed DM tools. While there's often some overlap in functionality, they aren't the same thing. For example, DG tools don't create data layouts or, in general, executables. Instead, they support the various artifacts and moving parts of a program. This means traditional DM functionality such as glossary development and sophisticated administration of rules and policies.[3]

Talk to a corporate controller or a manager of documents in any large organization. The tracking and maintenance of policies, rules, manuals, websites, and so on can be overwhelming and requires formal administration. After a short time, your DG program will be maintaining its own artifacts and will also start to deal with artifacts of all of the other data and information management efforts. The best examples of this are policies. In most organizations, you cannot swing the proverbial dead cat without hitting a policy. In my experience, the potential for administrative issues can threaten the vitality of your DG effort.

Since the first edition of this book, the technology market for support of DG has exploded. An entire book could be devoted to the topic. Tracking the comings and goings of specific vendors requires the skills of a play-by-play sportscaster. So, this edition will focus on a basic framework of features and functions. Tools will be categorized but identified no further. Essentially, the theme of using a checklist will continue. This chapter will offer the checklist for what you may want to manage your DG artifacts, and how to do it.

---

[2]It seems that as you get older you tend to get more candid. I suspect that it is part wisdom combined with diminishing patience from clearly seeing mortality. Either way, I am candid with clients and will be clear here—too many of the readers of this are reviewing this section before they have read other parts of the book. There is this mitochondrial, primitive urge to go buy stuff right away among data people. This is hard to fathom. It never works. Do not buy tools for DG until you know that program has some legs. Do not automate until you have something to automate.

[3]Ladley, John and First San Francisco Partners, TechTarget Blog, June 2016 https://searchdatamanagement.techtarget.com/feature/How-data-governance-software-helps-ensure-the-integrity-of-your-data.

If DG says, "we need a glossary," then a glossary capability or tool needs to be stood up, populated, used and managed by some area within the enterprise. If your regulatory situation calls for data lineage, then DG needs to ensure (and maybe operate) effective application of data lineage technology.

Types of tools

There are several categories of tools. There is overlap between them, and sometimes vendors are not clear as to which category they believe they belong to. But this is a start.

- Discovery—Technology that supports learning about your data—where it is, what it means, how it relates to other data. Some of these tools utilize artificial intelligence (AI) and are very sophisticated, and these capabilities will improve over time.
- Administration and directive—These tools support administration of DG artifacts and processes. An example is work flow and collaborative mechanisms which can be used to get agreement on definition of data elements. Also this category covers development and implementation of standards and policies, so it covers data classification as well.
- Data Efficacy—This category covers data quality and related tools. The accuracy of data can be handled by data quality profiling, but specific data control tools also fit into this category.
- Data Provisioning—Tools in this area address access, distribution, and publish and subscribe functions. Data obfuscation and masking, while a unique capability, could also be placed into this category.
- Life cycle management—Oversight of data life cycles is key to many organizations managing master data, reference data. Redundancy, obsolescence, and triviality (ROT) as well as data archiving are in this category. Also, data mapping of data is in this area.
- Metadata—Any tool that supports "data about our data" is in this category. Data modeling, analytical models, algorithms, data glossary, rules, documentation of metrics, in other words everything you need to know to make data management and governance work.
- Data storage—Given the enormous range of choices for sourcing and storing data, DG often needs to address tools rated to data at rest. So specialized databases and file managers, like Graph, can enter into DG conversations. Also, data acquisition, sharing, and selling fit into this area, as does cloud vs on-premise storage.
- Provenance—These tools support the understanding of the pedigree and interaction of your data with your organization. A tool in this category will feature lineage and impact analysis features.

> Since this book is limited to high-level coverage of this topic, here is another resource from a bit of research I did while at First San Francisco Partners. https://www.firstsanfranciscopartners. com/blog/category/data-governance-tools-and-software/.

Below, in Table 9.1, are a list of features, grouped ROUGHLY by how they tend to nestle in tools. Again, to dive into each area of functionality would create another book, so refer to this, as well as other references sprinkled in this section, to enhance your tool acquisition.

Table 9.1 Groupings of tool features

Administration and directives	Data agreements
	Regulatory compliance
	Data collaboration
	Policy management
	Rules
	DG Socialization
	Workflow
Data provisioning	Masking
	Data access
	Data access
	Obfuscation
	Visualization/dashboards
	Privacy/security
Data storage	Cloud
	Graph
Discovery	Lineage discovery
	Impact analysis
	Reverse-engineering data models
Efficacy	Data quality
	Data profiling
	Data controls
Life cycle management	ROT
	Data archival, destruction
	Data movement
	Data mapping
	Roles management
Metadata	Glossary
	Analytical models
	DG and DM metrics
	Data model
	Learning files for AI models
	Data catalog
Provenance	Data provenance/lineage
	Data location

Approach considerations

The key shapers of your tool strategy are

1. Readiness to automate DG capabilities
2. The scenario or use case you need to support
3. What your current tool stack looks like

I do not count the approval to get a tool, or procurement details as a shaper of DG tools approach; those are baseline activities for all types of tools. But you must understand what type of functionality you need, and make sure it fits into other DM functionality.

Readiness for tools

Your strategy work may have indicated some sort of role for tools. Before you proceed, you need to confirm your program will benefit from a tool, and you can effectively operate that tool. Here are a few scenarios to guide your thinking:

1. **Highly regulated industry**—Data lineage and discovery will support compliance. Obviously, metadata tools will document meaning. You still do not need to go buy tools until you know what you operating model looks like, but it will not be long before a tool will be most helpful.
2. **Master data initiatives**—A common, major data initiative is MDM. MDM flat out will not be sustainable, and therefore wastes a LOT of money, without DG. But supporting tools are not necessarily mandated until the DG activities are underway. Usually the MDM vendor supplies some sort of metadata. The useful metadata around MDM is often mapping old things to new. The master data should clear up semantic differences across business functions, so the need to manage common data definitions, standards, lineage, and reference data makes mapping and glossary type products handy.
3. **Advanced analytics/Big Data activity**—This is an interesting area, as a lot of benefit can come out of a data science area without any DG oversight at all, but *only to a point*. At some point a data scientist will say "we are getting slowed down by data quality." Or inconsistent definitions, etc. Quite often, the data scientists, while quite expert on statistical methods, have no clue about data management. I have had data scientists tell me that "there may be an issue with data quality here. Have you heard of this?" At this point they want to write their own tool, but data discovery and data quality tools and statistical model management enters the discussion instead (hopefully).
4. **Artificial Intelligence/Machine Learning**—Probably the only area where I will get keenly interested in tools well before other scenarios is AI. That is because AI, depending on the application of course, can go very well, or horribly wrong. And sometimes it is hard to tell the difference. Given distortions in AI based on model bias, data quality, and the operationalizing of erroneous models, AI often requires proactive data profiling, discovery, and significant understanding of data lineage.

If you think you need DG technology, make sure you can actually implement and support the tool. Even if you can identify with the above use cases, you also must ensure that your organization is ready to use a DG tool, as readiness is a huge factor in the decision-making process and the success of a DG program.

What should be tracked and managed?

Many tools try and do multiple functions, but, as of this writing, no single tools executes all of the various requirements we have covered at uniformly satisfactory levels.

A word of caution—there are no clear lines of demarcation between some DG tools and other categories, like data quality, data access, or DM. For example, there is a category of tools that have assembled the label of "Data Glossary." In general, they try to do a lot of things that ease out of the boundaries of a glossary. However, no matter what the software market, tool capabilities are always overstated by the vendors. Act accordingly. There can be a lot of overlap. Your choice in tools may be dictated to a huge degree by what you already own, i.e., sticking with a known vendor with several products already in house will be better than taking on and interfacing a new vendor. Also, SaaS needs to be considered—you could easily operate your DG support tools in the cloud, with no resident server or software. You truly need to do some serious consideration of a lot of factors.

All of these tools help you maintain a hierarchy, or taxonomy, of elements and artifacts that DG will need to consider, track, create, use, manage, and administer. There are also elements that DG will specify regarding tracking and use. In other words, both sides of the "V" have documents, policies, standards, and such that will require administration.

Business elements

The categories of business elements that are good to track are:

Business alignment—The business alignment elements are made up of documents and files that express business direction, performance, and measurement. These elements must be monitored by DG because they are the direct component of business alignment. As we have said often, ensuring business alignment to data asset management is a crucial activity for DG. These include:
- Strategy
- Goal
- Objective
- Plan
- Information levers

Business capabilities and processes—Business capabilities reflect the WHAT a business specifically does to operate and achieve its goals. Process elements are everything that has to do with events and actions that do something with data. If you have a process modeling tool, for example, the artifacts from this tool would be addressed in this area. From a DG standpoint, capability and process elements must be reviewed to ensure that controls are documented, as well as the key regulatory or compliance processes. Certain aspects of processes are important for data lineage and data provenance. Other processes, like events or communications, may require DG when the content for an external communication needs to be reviewed.
- Event
- Meeting
- Communication
- Training
- Process
- Workflow
- Life cycle
- Methodology
- Function

Policy—Policy elements have to do with artifacts that codify or document desired or required behavior. Obviously, DG will need access to these and, better yet, track them. A prime example is governance of documents, where there are legal, risk-based, and practical policies, which are often in conflict with each other. For example, everyone wants to keep those memos "just in case," while corporate counsel says to get rid of them as soon as possible. We include Principles in this category because policies stem from principles.

- Principles
- Policies
- Standards
- Controls
- Rule
- Regulation

Organization—This element covers the various roles and organization charts. DG will need awareness of this to manage who stakeholders and decision makers are. Granted, this is not an element you would need to place in an expensive tool. A spreadsheet would probably suffice with most organizations, but larger organizations may require a database of some sort, or use of the organizational entities in an enterprise modeling tool.

- Level
- Role (RACI)
- Location
- Assignment
- Community
- Department
- Team
- Roster
- Stakeholder
- Type
  - Steward
  - Custodian

Business data and information requirements—This element must be traceable by DG. This is because it is critical to ensuring business alignment. In addition, one area where organizations go awry in the information realm is the poor identification and tracking of requirements. A critical function of DG is to monitor and review the development of enterprise information management (EIM) requirements:

- Metric or measurement
- List or domain
- Event
- Subject

Permanent Artifacts—Documents or anything else that is stored permanently for subsequent use or review.

- Manuals
- Charters
- Presentations
- Work
- Project deliverables
- email
- Policy—written versions
- Principles—written versions

- Publications
- Website
- Work products from all EIM projects

Data—Knowing where data resides is important. So this element is not talking about the actual occurrences of real data, but rather where it is and what it means. The term metadata is used as well; however, that term is subject to overuse and distortion by vendors. This element represents all of the "data" required to be used by and to operate the DG program.

- Metrics
- Statistical models
- AI learning models
- Data Model
- Data Standards
- Dictionary and glossary
- Definitions
- Metadata
- Digital processes
- Scripts
- Programs
- Blog
- Wiki
- Files
- Business Information Requirement
- File/database
- Location

Technology—It is also good to track the technology that can use and affect data. This element represents the information about technology used to manage information assets. Honestly if you have CMDB or are standardizing based on ITIL, use those taxonomies.

- Product
- Hardware
- Software
- User

Ramification and benefits

There are many tools that can greatly enhance the operation and effectiveness of DG. There is no one tool that can be used to track all of the elements scrutinized by a DG program. Therefore, you need to think carefully about how you will track the various types of artifacts and automate them as efficiently as you can. Too many tools will present expensive cost of ownership issues.

You need to be cautious about throwing up myriad Excel spreadsheets, since these can become unmanageable. The best approach is to use as many existing tools in the information management (IM) realm as possible—that is, the modeling tools, the enterprise architecture tools, and various catalogs and productivity tools. These operating costs are already absorbed to some degree. Then enhance your DG with additional features, like lineage.

SharePoint is an excellent option for linking and tracking myriad objects, but *only if designed and used efficiently*. Using SharePoint as a dumping ground for artifacts is useless and costly. Using Wikis as an internal entry point is done with great success in many organizations.

Whatever the technology, try and use a central entry point. This is where SharePoint or a Wiki are beneficial. At its core, DG is a program that sits on top of a defined set of rules and workflows. (This is yet another reminder that DG is not special; it is just another business function with the same operational requirements.)

In DG, technology is a challenge in developing workflow and document management. Taking out lists of elements and creating an internal taxonomy for DG is an ideal approach.

Workflows that you may want to wrap around your tools, or evaluate tools on, are core DG events. Tracking all of the items in taxonomy is ideal. For example:

- Exemption requests
- Changes to standards
- DG issue resolution; unresolved items

These are all DG processes that can be adapted to workflow and document management.

A few other technology components that can help DG are chat forums, data control products, and policy management tools.

Chat forums are useful as outlets for questions and advice while capturing crucial feedback on feelings about DG. Data control products have been around for years and offer excellent facilities to enter and observe the execution of rules. Policy management tools have also been around and offer a set of tools for DG stakeholders.

The bottom line for DG technology is that you will be striving to assemble a set of technologies. Many vendors, as of the writing of this book, are moving into the area of pure DG administration. Some are entering from the IM realm, others from the document management realm. Regardless, you will need to assemble a toolbox of capabilities. Create your own internal taxonomy (which can be woven into a taxonomy tool) and connect the taxonomy via a single-entry point to your other tools. Again, traditional places to store artifacts, like SharePoint and Excel, are useful, but only if managed and, well—duh—governed.

Operating frameworks

Normal people... believe that if it ain't broke, don't fix it. Engineers believe that if it ain't broke, it doesn't have enough features yet.
Scott Adams

This area of design activity is applied when you need to get detailed on the HOW DG will work. A series of activities will produce processes, roles, responsibilities, and operating models. In addition, you can apply a concept from the software industry, that of a minimum sustainable model, to create a baseline DG program that can hold its own regardless of changes in your environment. You are doing some engineering here. Applying the outline supplied by capabilities and inserting detailed features to ensure the capabilities can function.

If you chose to, or need to, do all of the activities in this section, you will identify processes to fulfill the capabilities, apply who is accountable and responsible, define a framework that will support the operations of DG, and determine what the absolute minimum sustainable program looks like (Fig. 9.9).

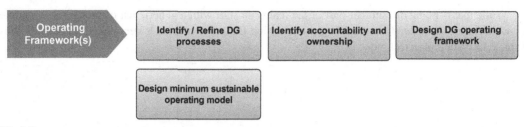

FIG. 9.9

Operating framework(s) activity

Identify/refine DG processes

Any program has a point where the ideas, concepts, and philosophy must become real, tangible, and actionable. This activity is that point for DG. The mission and vision of DG, along with the principles, business drivers, and capabilities, converge to identify the policies that codify DG as well as the actual processes that will be required for a functional DG program.

Some of the processes for DG will have the weight of carrying out policy, so be aware that policy and process are not mutually exclusive. Other processes will make sure the activities of DG are carried out. For example, there will be functions to determine or revise data policies, and there will be functions to audit and verify compliance with data policies. Other process will simply be decomposition of your required capabilities, such as supporting analytics or data quality.

Another consideration during this activity is the development of the processes (and process or workflows) for managing the artifacts and outputs of the DG processes. These essential processes, such as issue resolution, need to be detailed. Policy maintenance can be overwhelming in a large organization, so that should be addressed as well.

Approach considerations

The actual activity occurs along parallel efforts. Fig. 9.10 shows how the principle provides input and inspiration for other components of DG. First, the information principles need to be evaluated for implied policies. Review the "Authoritative" principle we displayed earlier. Each implication points to a potential policy, that is, how to you make sure you have dealt with each implication?

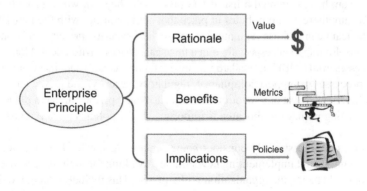

FIG. 9.10

Principles imply policies and value

While that is going on, another part of the team can start with a generic list of processes (a sample of which is in the appendices) and begin to develop the process list. Or decompose the capabilities (Fig. 9.11), a common technique.

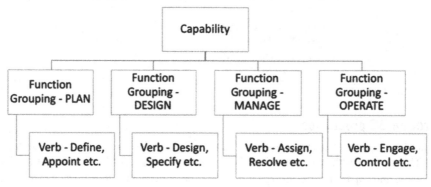

FIG. 9.11

Capabilities related to process

Principles (specifically the implications) are used as a basis for initial identification of policies. The team will merge these with a generic set of processes typical to a DG program. The essential DG processes need to address a basic functional cycle of *plan, design, manage,operate, and sustain* and cover processes to:

- Sustain key business measures or metrics model
- Support standards, controls, and policy
- Support master data and enterprise resource planning (ERP) projects
- Support Big Data, and advanced analytics
- Support real AI and machine learning
- Support regulatory drivers
- Manage enterprise data model standards and procedures
- Manage processes for reference and code data
- Plan and manage the DG program itself, including processes to administer policies and standards

Once the DG team has assembled a list of DG processes, they can work through a rationalization process and make sure there are no policies in place that could conflict with the new DG processes.

Separately, the team can address regulatory items such as security, privacy, and compliance.

Data controls are also important, especially if in a financial services environment. Many organizations derive their DG processes from COBIT, a standard framework for data control and financial governance.

Don't forget processes to support compliance, regulatory, security, and privacy areas. These will be very visible DG functions. Most organizations have security and privacy areas in place so coordination and leverage of existing policies in this area is important. If the policies already exist, make sure they are adopted into DG.

Lastly, the team needs to strongly consider some process design for key functions such as issue resolution or maintaining and implementing new policies. Walking through these will give a great indication of the amount of change the organization will undergo. This includes change to the development methods used by IT to implement systems. These are referred to as *system development life cycle*

(SDLC) *methods* and can take a variety of forms (Agile, waterfall, iterative, etc.). Regardless of form, the DG program will require changes to internal SDLCs.

The first edition provided task lists for every activity, and they still appear in the appendix. However, this particular activity requires we review some of the details without flipping to the back of the book. You may want to do some or all of these, depending on your approach. A low-profile effort may want to consider these steps and select a few. A larger effort may want to seriously consider how to approach all of the potential process design. It may even be a set of activities for roadmap increments, with new processes rolling out as new capabilities roll out. Defining functions and processes can be enabled by breaking down capabilities into functions—a pretty typical processes design technique.

Steps to consider when defining processes:

1. Gather any existing information and governance policies.
2. Identify processes to sustain key business measures or metrics model.
3. Identify processes to support standards, controls, and policy.
4. Identify processes to support master data and ERP projects.
5. Define/support regulatory drivers.
6. Identify requirements and processes for enterprise data model standards and procedures.
7. Identify requirements and processes for reference and code policies/procedures.
8. Identify any organization periodic strategy, planning, or management functions where DG can participate.
9. Identify processes to administer policies and standards.
10. Optional: Work with finance and compliance:
    (a) Identify gaps in current state of data management.
    (b) Specify adequate controls.
    (c) Specify privacy and security concerns.
    (d) Specify compliance and regulatory concerns.
11. Specify key DG process flows:
    (a) Define issue resolution process.
    (b) Define process for DG policy and standards changes.
    (c) Define DG and project interaction.
12. Develop new organization performance objectives.
13. Identify other DG detail processes and other areas where DG will affect development or "time-to-market" type processes:
    (a) Identify changes to SDLC, Agile, etc. processes.
    (b) Design DG process details, deliverables, documentation for SDLC integration touch points.
    (c) Develop revised process/policy alignment plan (Review/update existing policies and processes related to DG and EIM)
14. Ensure processes and policies are not in conflict.

Ramification and benefits

The primary benefit is the ability to see the activity that is required to operate the DG program. Of course, the downside could also be that the business gets to see what is required to operate the DG program! The bottom line is you have moved from abstraction to reality. At this point, especially with higher exposure efforts someone will see that a big DG program can be a lot of work. This is normal. A large DG program IS A LOT OF WORK. But that is addressed in how you deploy the program.

You cannot shave off required processes and policies. In the first edition we did the operating model before this step. I have reversed this, so it is easier to build an incremental operating model to accompany incremental process roll out.

If you receive feedback upon review of your functional design that it is "too much" remind the critics that business functional areas have similar sets of activity. And of course, the functions will roll out incrementally, never all at once.

Remember, you must identify the DM/IM functions as well as the DG functions. This is done to make a clear distinction between the governed and the governors. Business areas do not have a difficult time understanding they are subject to oversight. Business leaders often interact with all kinds of compliance requirements. But IT staff as well as information managers of an organization occasionally have difficulty seeing the distinction. Sorting out the DG processes added details to the left part of the V. Don't forget the details for the right side.

Separation of duties is an important concept. If the same people handling DG activity must also maintain databases and manage data models, then you do not have proper oversight. There will be an inevitable conflict between the projects they are assigned and the governance of those very projects. The same goes for the business sponsors and stakeholders of projects. They cannot be expected to be motivated by project deadlines and then stop and audit compliance to governance policy. Inevitably, the governance falls by the wayside. Make sure you provide examples that show how DG and IM functions operate independently and together.

Sample output

A generic list of basic DG and DM processes is in the appendix. Sorry to force you to the back of the book, but space is limited for these larger lists. Figure 9.12 shows how functions will lay onto the "V"

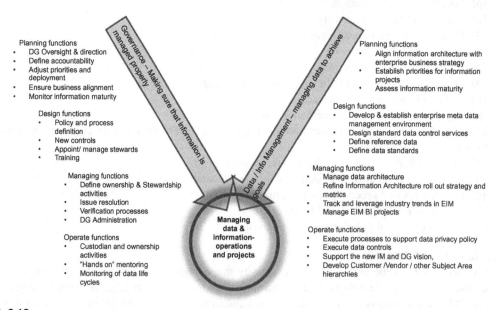

FIG. 9.12

"V" Model with functionality

to demonstrate on which side activities land. Review this list of possible outputs and you may see something you need to add to your effort, regardless of the approach:

- Draft DG policies
- Metrics and Business Information Requirements management processes
- Standards and controls management processes
- MDM and ERP DG processes
- Analytics and Big Data oversight processes
- Regulatory DG processes
- Data standards DG processes
  - DG planning and management processes
  - DG administration processes
  - Policy/process cross reference
  - Reference and code DG processes
- Processes to close current DG deficiencies
- Data controls
- Privacy/security controls
- Compliance and regulatory DG processes
- DG Issue resolution flow
- Policy and standards maintenance flow
- Project DG flow
- DG performance objectives for business areas
- SDLC change requirements
- SDLC changes
- Revised polices affected by governance

Identify accountability and ownership

This activity adds details to the DG functions and processes, if necessary, of "who is responsible." You are not naming names, per se, although the pressure for names will start immediately. You are developing the view of who will perform the various roles in the organization after DG is deployed, but only generically, by department or position. This will lead to development of the various layers of authority within operational DG. This has to start with a view of where accountabilities and responsibilities lie. By this point, even if you are low profile, someone will be asking "who is responsible?"and"is this a new job or do I need to make time for it?" At some point, the appointment and steward-like roles are required before you assign titles like "steward," or "custodian."

Approach considerations

Focus on processes where DG will touch the business in the form of an individual being accountable for DG success, e.g., oversight of stewards. Along the way, look for responsibilities vs accountabilities. Apply a reasoned process to show the organization the most desirable framework for DG. For a low-profile effort, you will need to focus on ONLY the immediate capabilities and the related processes.

Start like any other design—what is required? This come from the capabilities and identifying processes to satisfy the capabilities. Using a list of processes as we mentioned, or even doing one of your own from scratch (but why?) will give you a functional design. Remember this may be a lot, especially if your program is being driven by a bunch of initiatives, like Rocky Energy.

Once you have a crude functional model for DG, some sort of identification of responsibility and accountability is necessary. For a low-profile effort, this is usually some new responsibilities and the sponsor is the gatekeeper for accountability. It will stay that way until the early efforts start to become embedded as regular operations.

For larger efforts develop a RACI chart. This means scrutinizing all activities described as necessary to perform DG and IM, even if their implementation is far away.

If you are engaged in a large enough effort to do a RACI chart, it will take time. There will be multiple passes at this work product and much debate. Most likely, there will be an issue or two arising from this process that will require steering committee or executive-level intervention. Should the politics be too intense, wait until you have the operating models in hand before socializing new accountabilities. Showing leadership that you are not creating new empires and departments, just assigning new behaviors, will ease concerns.

Sample output

Fig. 9.13 represents a sample of results from a RACI analysis.

Given the potential for the DG team getting the first real dose of resistance (usually in the form of the organization expressing concern that this is the correct thing to do), this should not be a set of linear tasks that are executed without external contribution. Every output from this activity requires continuous vetting with potential stakeholders. Every step requires sensitivity to culture and politics. This does not mean the DG team dumbs down what governance is supposed to accomplish. It means to be resolute and navigate through the first set of real barriers.

Have the DG team address this work product before any discussions or reviews are held with other stakeholders. This will help the team tighten any loose ends and prepare explanations of the meaning of certain processes.

Ramification and benefits

Business ramifications can be high in this activity. Anytime new responsibility or accountability enters the picture, organizations react. The reactions can range from a few raised eyebrows to a request for a full-blown human resources engagement.

The benefit of this activity is the raising of organizational flags at a relatively early time. Once leadership starts to digest the accountability and responsibility inherent in DG, then you will know where they really stand and where your support is located.

Even a low-profile effort can take 2 weeks to sort out an operating model, and there may be more time involved. It may take 2 weeks to do the first cut at the chart, but it may take another 2–4 weeks of walking it around to make sure you have support. Remember that there is high potential for politics and resistance as the DG team gets closer to the reality of implementation.

Larger efforts most likely will incorporate the RACI analysis into the change management efforts, which needs to be a parallel effort at this point.

Management Phase	Basic Enterprise Information Management & Information Governance Functions	Enterprise Info Mgmt Functions	Exec Steering Committee	Data Governance Forums	Data Governance Council	Organization Change Management	Project IM Functions	Project Mgmt Office	Project Steering Bodies
Plan	Align applications and project with business	C					A		
	Share and educate business area and projects on IM and DG Policy and direction	A,R			C	I	I	I	I
	Plan principles, policies, standards, and controls for enterprise governance			R	A		R		
Define	Identify gaps and refine enterprise IM road map and environment	A					R,I		
	Identify gaps and refine enterprise IM roles, processes and metrics	A					R,I		
	Identify gaps and refine enterprise IM principles, policies, standards, and controls	R		R	A		R		
	Define new principle, Policy, standard or control / Change to existing principle policy, standard or control	R	A			I	C		
	Define enterprise BI and reporting matrics	A					R		
	Define processes to identify certified information sources	A			I		R		
	Identify certified sources of information	A			I		R		
	Define enterprise IM and DG Organizations	A					R		
	Acquire new tools	C,R					R		A
Manage	Approve enterprise IM principles, policies, standards, and controls	I	A	I,C	R	I	C		
	Define enterprise IM Organization change strategy	A,R				A	R		
	Manage information architecture, incl. data models, canonical models, rules & definitions, meta data	R					A		
	Manage information portfolio (actual files, data bases, content, data stores for ACME)						I,R		
	Follow existing principle, policy, standard, or control	R	A	R			R	A	
	Develop project documentation	C	I				R,C	A	
	Maintain inventory of certified information sources	A		R	A		R		
	Support and facilitate custodians	C					R	A	
	Ensure sustainable application data quality	A					R		
	Support and use enterprise IM and DG technology (Repositories, models, DQ tools)	A	A		R		R	I,C	
	Manage and resolve DG and IM issues	C		C,I	A		C	C	A
	Enforce enterprise IM principles, policies, standards, and controls	R		R	R		I	R	
Sustain	Ensure enterprise DG program is followed	R	A	R	R		R	I	
	Execute culture change management methodology tasks	I	I,C			A			
	Measure people performance to enterprise DG and IM goals					A			
	Measure progress towards enterprise change goals					A			
	Execute Communications Plan	C				R			A
	Execute Training and Education Plan	C				R			A

FIG. 9.13

Sample RACI

Sample output

See Figs. 9.14–9.16 for the results of this activity and the next.

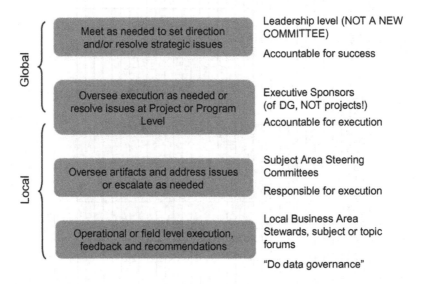

FIG. 9.14

Operating framework levels

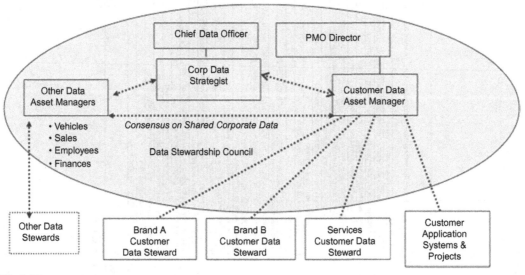

FIG. 9.15

Federated stewardship

Activities

- DG strategic roadmap and capabilities
- Budgetary Support

- Charter
- Road Map Oversight
- Technology Strategy
- Sustaining Activity
- Policy Management
- DG Metrics
- Policies

- Processes
- Standard
- Definitions
- DQ Metrics & KPIs
- Issues Log
- Data Profiling Scorecards

Strategic Guidance

Oversight and Management

Executive Oversight (Sponsors)

Data Governance Steering Committee

Data Governance Working Group

Data Project Oversight (Business)

Data Project (IT Data Lead)

Tactical/Execution

Cross-Functional Business Data Stewards

Product Data Custodian

Inventory Data Custodian

Operations Data Custodian

IT Enablement

Issue Escalation

Roles

DG Oversight:

- Propose adoption of existing committee, else create net-new Sponsors/Steering Committee
- Agree to Data Governance Strategy
- Alignment on Program Scope & Roadmap
- Ultimate adjudication point

DG Execution

- Profile data domains
- Adjust projects
- Remediate data issues
- Provide measures of progress and value

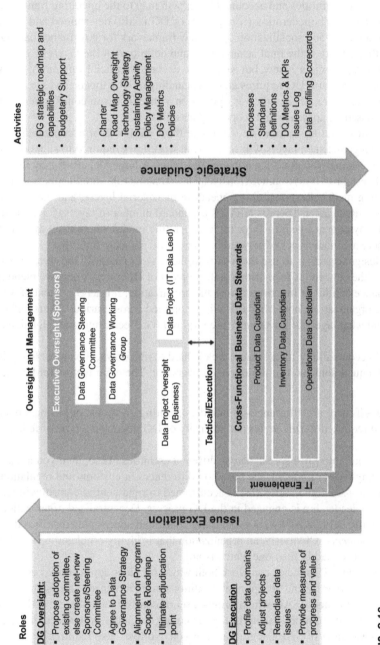

FIG. 9.16

Sample operating model

Design DG operating framework

This activity refines the functional framework from above and defines the operational layers required to do governance. The responsibilities and accountabilities will map to the operating framework and may require a few adjustments. The federation (if required) of DG is also determined and specified. After that, the process of approval and socialization of the new framework for governance can start.

At the core of this activity is the final arrangement and definition of the layers of governance. The prior activity implied a preliminary view, but we need to declare an official version. Again, there is not a single formula, and successful DG requires an understanding of why this is the case.

As always, these items are all presented as options on a checklist. You could combine this activity with the prior one, based on approach and scope of your effort.

The layers in your operating framework will come from the levels of accountability and responsibility. The interchange between layers will result from workflows required to support your required processes. This is a task that you don't want to take to excess. A large DG effort may require all kinds of processes to fully establish its vision. Obviously, an accountable role should appear at the leadership layers in an operating model. Responsibility appears in managerial layers, and the execution of DG activity is in an operational layer. There is NEVER a standard number of layers (for some reason there is a tendency for many clients I visit to show me an operating model where everything is forced into three layers and there are many types of stewards). There is no mandate for either.

An important design aspects is to use the results of the RACI to identify the style or nature of federation. Remember, the concept of federation (in the context of DG) means how we blend and stratify the various governance entities or functions across the organization. It is a refinement of where the DG elements touch the organization, how standards will be applied across various layers and segments of an organization, and what layers of governance are required (i.e., local, regional, global, enterprise, or others).

For example, if accountability for a subject area is hard to nail down, then most likely it is used in a context that will require some sort of multilayered oversight. The main factors for how federation is established are:

- *Enterprise size*—If there are differences in brands, operating divisions, or business operating models that will require differing styles and intensity of DG, then some type of federation needs to be defined.
- *Geography*—Is your enterprise spread across different countries? If so, then you are almost guaranteed varying types of governance based on differences in customs and regulations.
- *Organization style*—An organization that is accustomed to rigid central control will tend to adapt easily to DG, *if* its leadership is engaged in the DG process. Decentralized organizations will require *very* specific definition of what is centrally controlled and what is distributed.
- *Regulatory environment*—Obviously, an organization that is highly regulated will embrace central control of assets more readily than one which is not.
- *IT portfolio condition*—This factor can work both ways. An older application portfolio can create a desire to build anew and accept new conditions of governance. This is most common when a company implements SAP, which brings a set of constraints that are mostly based around success factors. Modifying functionality in SAP is not a good idea—you accept it "vanilla." Sustaining the advantages of SAP integration after you "go live" also requires ongoing DG. The configurability of SAP can allow users to run amok. It is not uncommon to find SAP master

files as badly managed as the legacy files they replaced.[4] Conversely, a beloved, embedded (or tolerated) legacy application can be a barrier. It is considered either ungovernable or immune from any perception of disruption. Lastly, if you combine a geographically dispersed company with a diverse and eclectic blend of applications, any kind of federation on a central basis is going to be an architectural challenge.

- *Enterprise architecture*—This factor is difficult, because it cannot be changed very easily, if at all. The symptoms of an eclectic application portfolio and inconsistent and unplanned enterprise architecture produce the same challenges that create the need for DG. There is an entire other book to be written on the role of enterprise architecture and information asset management (IAM). So, briefly, enterprise architecture (the blend of all of the elements of People, Process, and Technology) or EA, can really influence federation. An organization with no formal approach to managing the blend of People, Process, and Technology will need to be almost militant in defining some sort of central DG. This is because the DG program, for right or wrong, will be taking up the slack due to poor technology governance. An organization with a decent or robust approach to EA can leverage the dickens out of its IT and technology governance and define very clear lines of federation.
- *Culture*—The cultural factor can be divided into two subtopics, maturity (we called it IMM, or information management maturity) and capacity to change.
  - *IMM*—If an organization is not mature in terms of its understanding of information usage or handling of its information assets, then federation should lean to more rigorous or centralized. Of course, the immaturity will have resulted in a lot of informal information assets scattered about.
  - *Capacity to change*—DG means change. Many types of organizations are unaccustomed to or in denial of the need for change. Older cultures or closely held companies typically have lower capacity to change, while younger organizations may be more amenable (but not necessarily).

All of these factors must be blended to determine the type of federation required to carry the DG processes forward.

Federation then needs to be combined with the various layers of governance that will evolve from an analysis of the RACI chart. For example, if we determine that customer data needs to be governed centrally, but the applications that use customer data are scattered about the globe, the accountable and responsible parties will need to be identified with some consideration of the distribution of authority. Therefore, there will need to be a centralized flavor of customer DG as well as a distributed flavor, and a collaborative set of processes will be required to facilitate DG for the customer subject area.

Of course, the framework to manage the various striations of governance will need to consider the federation and layers of DG. Remember, there is no independent organization chart, so you are weaving DG within the existing organization chart.

---

[4]The author got into trouble years ago after writing an article describing SAP software as "instant legacy—just add money." SAP took great offense to this, but they missed the meaning. If you treat the SAP application data the same as you treated your old systems data, you get the same result—junk data. At an average cost of $35 million per project (author's data), that makes for very disappointed CEOs.

The blend of federation and layers of DG oversight produce the representation of operating layers and federation, usually in the form of a hierarchy or network.

Approach considerations

Remember that the RACI is the main input into this activity. You could do it without a RACI exercise if your efforts are low profile and with obvious resource roles and accountabilities. Just remember you will then need to revisit this activity later as your DG program expands. Eventually, you WILL need to engineer an operating model and be able to point to some formal method to justify your recommendations. This activity almost always will appear with the prior activity of identifying accountability and ownership.

Any stakeholders who are accountable need to be positioned in the operating framework in such a manner to observe and communicate cross functionally.

It is common to think in terms of three layers for operating frameworks, but I find that self-limiting. Fig. 9.13 shows you need to blend global and local practice with strategy, tactics, and execution. If your organization is geographically close, and has a typical organization hierarchy, then your operating framework might be three layers. For large organizations you can easily have four layers and could conceivably have slightly different execution layers within operating frameworks based on the domain or discipline being governed (Fig. 9.14).

Your model is determined by the necessary functions or processes, and where they are applied to various domains. Fig. 9.15 show a large organization with a highly federated model. (Also note the steward in a position of accountability in this example.)

Fig. 9.16 is a small organization, that started "noninvasively" but quickly realized its culture would not support anything new unless it came from the top. Since it is a small organization, the various strategic and tactical functions could be handled by only two layers. Federation is only by subject or domain, vs Fig. 9.15, which is federated by domain, brand, and applications. Thanks to First San Francisco Partners for the permission to use this redacted example from work I did while at that organization.

Ramification and benefits

Regardless of your approach, you will have an operating model. Even the noninvasive efforts needs to clearly depict how DG will work. Remember that it takes a long time for leadership to actually understand what DG is, even after it has started. Everything you can do to make it clear is crucial. For example, a low-profile effort has a "steward" correcting reference data. This seems innocuous but leadership needs to know there is a new job responsibility.

Sample output

Fig. 9.15 shows a federation example of a multinational, multibrand company.

Fig. 9.16 is an operating model from a small, one location, one business model, organization.

Design minimum sustainable operating model

This is an entirely new activity. That is, it was not in the first edition. While at First San Francisco Partners (FSFP) the DG practice incorporated a concept from the software industry—that of a minimum viable state, or MVS.

"The term Minimum Viable Product, with its roots in the software industry, has been part of the lexicon for some time now. Its meaning: something that provides the most basic functionality and value to satisfy early adopters. There's a similar concept in First San Francisco Partners methodology and approach, the Minimum Viable State (MVS).

The MVS is a step toward the future state and is less mature than a recommended state, because the recommended state would provide more direct and measurable business value.

The MVS is truly that *minimum* demonstration of value so key stakeholders can agree that the data program is a solid, strategic idea and one they will allocate money or resources to move it forward."[5]

Before you drop the book and go rushing off to define your MVS, please consider these key points:

1. It is a minimal statement of value. Not much different from the first noninvasive effort, or a proof of concept. It is an INTERIM step to a longer-term goal.
2. It depends heavily on consumer acceptance. Like Minimum Viable Product (MVP), it requires acceptance and delivers basic functionality.
3. It still requires an operating framework or model, that is, it requires some sort of thought. In larger organizations, MVS can require significant engagement from stakeholders to define exactly what the minimal value of DG should be.
4. MVS means that you need to define and deliver a minimal set of functionalities. But MVP and MVS are based on software and consumers. It is a pull model. The consumer must accept and embrace the product. And once you get a customer, you have them. DG is a push model—most of the time you are implementing a capability that most stakeholders are not sure of. They do not have the motivation of a consumer. So MVS is a good approach only if the operating model is compelling and sustainable.

Hence, we need a minimum sustainable operating model (MSOM).[6] What set of roles, communications, and workflow is required to keep DG in operation, displaying a level of minimally required value yet being resistant to resistance to change and the ebb and flow of organization and business conditions? MSOM could end up being the essential operating model for a long time. How many initiatives get started and take a longer time to flourish than originally planned?

Table 9.2 shows some possible characteristics the MSOM may display—specific capabilities and processes, or a reactive "problem solver" presence, vs a proactive oversight presence.

Approach considerations

Almost every DG engagement could benefit from considering an interim, sustainable, operating model. Leadership may challenge your program and say "minimal sustaining operations will be fine forever." That could be true but remember it will not deliver the full value planned for DG. It needs to be made clear that this model serves only to prove value and embed operational DG in the organization to some extent. If your first step is a noninvasive approach, but oriented toward a POC, it MAY NOT BE MSOM. Your MSOM in a noninvasive approach may FOLLOW the first iteration or two.

---

[5] A Minimum Viable State and Why Information Management Programs Need It, O'Neal, Kelle, 2018, from the FSFP Blog, https://www.firstsanfranciscopartners.com/blog/minimum-viable-state-information-management/.
[6] I am not a fan of lots of acronyms. MSOM is not meant to become an industry standard acronym, like IBM™ or DBMS or IT. I just don't like typing "minimum sustainable operating model" over and over.

Table 9.2 Types of data governance presence

Types of DG functions that might "stick"	Data governance presence	
	Reactive	**Proactive**
Support planning	DG reviews strategic plans and provides impact analysis	DG participates in annual portfolio planning
Support design and architecture	DG review enterprise architects' output	DG supplies standards to enterprise architecture
Oversee DM	DG implements data issue report and resolution systems	DG creates and oversees implementation of data quality standards
Operate DG	Manage report request process	Oversee data access standards implementation

The MSOM should only have a few DG functions or processes. It can be based on one or two principles and policies. The key input to this activity is your long-term operating model. The key steps for defining the MSOM are:

1. Define the scope of MSOM if different from the long-term program scope.
2. Extract the necessary capabilities and processes that you want to make persistent as part of the MSOM. (This may require some facilitated meetings with stakeholders.) Identify what functions need to be visible and can be persistent. Refer to Fig. 9.6 again - the "bare minimum" column.
3. Verify and state slice of DG capability that can embed itself deep into operations. Where can the most value be added with minimal functionality?
4. Develop a roll-out plan (this can also go into your roadmap). But often the MSOM gets implemented in parallel to the completion of the long-term roadmap.

Ramification and benefits
Think of the MSOM as a first iteration of real DG. After a low-profile effort gets started, it still requires permanence. A larger footprint DG program also requires permanence with the first iteration. Once there is a permanent structure in place, you have a program that is in growth mode vs development. That is a substantial win.

You can target your culture (see Table 9.2 where you can be reactive or proactive). So you can add value while, temporarily, skirting around difficult resistance. This is why the recommendation to insert DG into planning is so common. I have seen companies place DG into the annual planning processes ONLY, as MSOM.

In the event of changes in your organization or environment, you still have a permanent structure that can be used to address whatever challenges are presented.

Sample output
The best way to show MSOM is to compare one to the eventual long-term model. Here is the MSOM for Rocky Energy compared to the long-term operating model (Fig. 9.17).

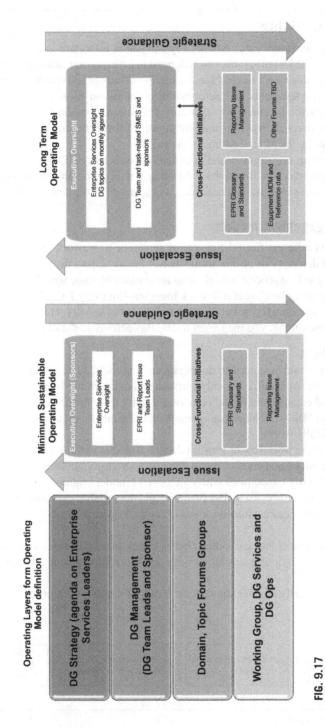

FIG. 9.17

Sample operating model

Engagement and workflow

The operating framework and model are certainly key, but just as important is describing what happens between the layers. For some DG programs you will need to go into detail about how DG engages with specific areas. Obviously, this won't be the case for smaller efforts or low-profile approaches.

You will need to complete roles and responsibility details on most efforts, even low profile. Remember, you want a permanent change of some sort. That means someone is doing something different. Even a noninvasive approach, where you might create a formal role from an informal role, means describing and documenting the official nature of the role.

Lastly, all of the operating, engagement, and roles need to be socialized. Leadership needs to hear what is planned. Don't go any farther into roadmaps or building training programs or doing project plans unless you make sure there is awareness, understanding, and approval of the new workflows and processes.

Design required engagement models

An engagement model is different from the operating model or framework. It is developed when you need to get into detailed descriptions of how things need to work, especially interactions between various functional areas. IT can be a process model, or an abstraction of some sort. Format is not important. What is important is depicting how areas will work together. Not every DG effort will need to do this. But if you have some critical interaction between areas that do not usually interact or need to learn how to interact when DG comes along, you may want to add this activity to the checklist.

Approach considerations

This paragraph or two assumes you need to do engagement models. Low profile, or noninvasive may not need to go into this level of detail, especially early on in the program. Mid-size efforts may combine operating framework and engagement models into one activity that describes DG operation.

If you are defining an MSOM, then an engagement model might be required for stakeholders to see clearly how DG will initially work.

To be clear—this is basic workflow or process design. There is nothing new here in terms of technique. Use whatever technique and level of detail that will work for your situation.

The first step is to identify capabilities with engagement model requirements. For example, a DG program may want to focus on supporting data planning and data quality capabilities. Other capabilities like data standards may not require extensive cross functional operations, so don't require an engagement model.

Next identify all participants, not individuals but business areas. Then describe how the work will flow (DG work) across the various areas along with any work products or artifacts that need to be handed off.

Lastly, make sure you verify the workflow to the RACI or similar work products to ensure consistency (Fig. 9.18).

| Engagement and Workflow | Design required engagement models | Complete roles and responsibility identification | Socialize operating framework, engagement models, and workflow |

FIG. 9.18

Engagement and Workflow activity

Ramification and benefits

This exercise can go a long way toward showing DG has relevance and can fit into an organization. If you consider that the ultimate goal of DG is to disappear into a business-as-usual state, it makes sense incorporate DG into important workflows, like Planning, and applications development. The two examples are from the Rocky Energy case study, Figs. 9.19 and 9.20.

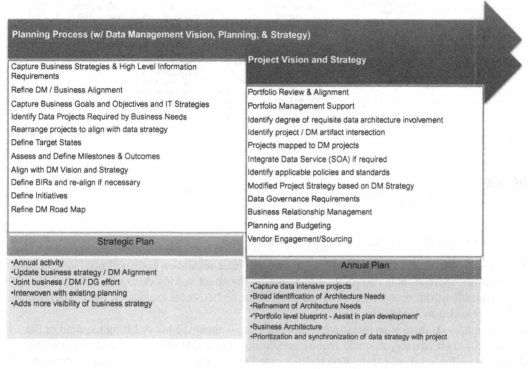

FIG. 9.19

Sample engagement model

Sample output

The engagement models below show a simple depiction of how Rocky Energy will overlay DG processes over existing capabilities, in this case annual planning, and the SDLC applications development cycle.

Complete roles and responsibility identifications

At some point names need to appear alongside the levels of responsibility and processes. They may be your current executive sponsors or DG team leaders and other participants at the beginning. If not, then there will need to be some initial socialization of the DG framework and vision for these folks.

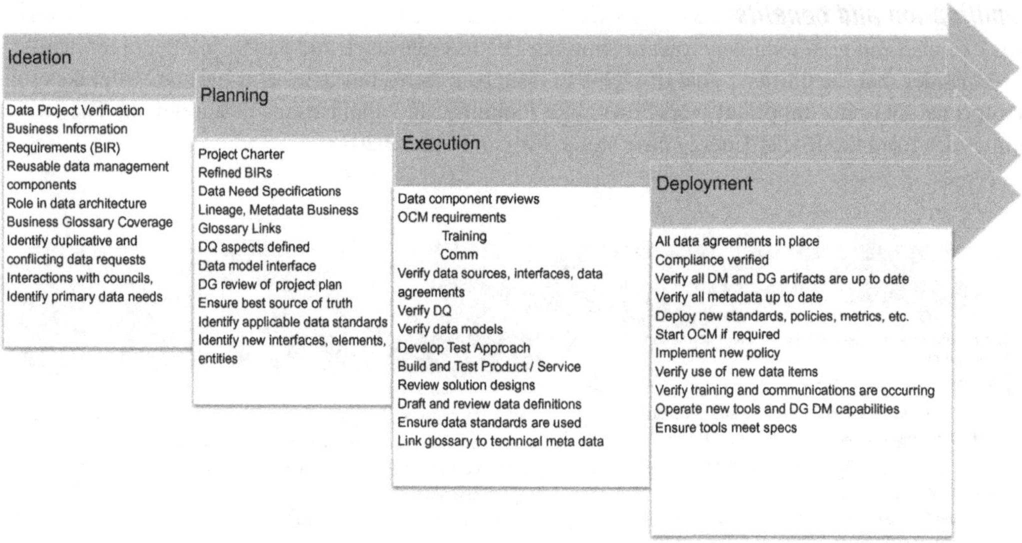

Ideation

Data Project Verification
Business Information
Requirements (BIR)
Reusable data management
components
Role in data architecture
Business Glossary Coverage
Identify duplicative and
conflicting data requests
Interactions with councils,
Identify primary data needs

Planning

Project Charter
Refined BIRs
Data Need Specifications
Lineage, Metadata Business
Glossary Links
DQ aspects defined
Data model interface
DG review of project plan
Ensure best source of truth
Identify applicable data standards
Identify new interfaces, elements,
entities

Execution

Data component reviews
OCM requirements
    Training
    Comm
Verify data sources, interfaces, data
agreements
Verify DQ
Verify data models
Develop Test Approach
Build and Test Product / Service
Review solution designs
Draft and review data definitions
Ensure data standards are used
Link glossary to technical meta data

Deployment

All data agreements in place
Compliance verified
Verify all DM and DG artifacts are up to date
Verify all metadata up to date
Deploy new standards, policies, metrics, etc.
Start OCM if required
Implement new policy
Verify use of new data items
Verify training and communications are occurring
Operate new tools and DG DM capabilities
Ensure tools meet specs

FIG. 9.20

Sample engagement model

Some of the roles and faces you will need to define are:

1. *Council*—Members of the primary monitoring and issue resolution body will need to understand their role. Individuals in this position must not be shy about making decisions. In larger organizations, this group will not be made up of the highest-level executives, but of staff that are well regarded by leadership.
2. *Committee*—If there is an executive committee (i.e., without the heavy lifting required of the council), their advisory role will require appointing individuals who understand DG and IAM.
3. *Forums*—These subunits that are topically focused require the same considerations as council and stewardship members. They are subsets of councils, but must be willing to dig into a specific issue.
4. *Accountable Stewards/Owners*—These appointees need to understand that they are information executives—and they must take the role seriously. They will be ensuring that DG as a mindset actually starts to "stick." If the information area for which they are accountable goes awry, they must be the right person to accept accountability, push an issue up to a council, or take action with subordinates when policies require enforcement.
5. *Non-Accountable Stewards/Custodians/Owners*—The stewards or custodians who are responsible, but not accountable, also need to accept a role that requires them to point out standards violations. Often these same personnel are participating in IM and development. They are at the bottom of the V.

This activity also requires the assignments of recommended staff be presented and approved. Lastly, and probably most overlooked, is the need to draft succinct charters for the various layers of the DG governing framework. An outline of a typical charter is in the appendices. Vetting the roles and charter is not a bad idea at this point.

Approach considerations

For a low-profile DG effort, you are really just appointing the permanent participants in DG, usually someone with a stewardship or custodial role. There may be more political problems than finding the right person.

A common mistake is to fail to keep the design of data and information management processes separate from DG processes. Don't forget to review data or information management processes as well. (That is why they are in the big list in the appendix.) This is usually because the initial staffing of DG is drawn from information areas. Often the initial DG staff are told to "fit it in" to their current roles and responsibilities.[7] It is challenging to these individuals to maintain the separation of duties mandated by the V, while creating embedded organizational processes, new roles, and a sustainable program.

Before we delve into the details, take time to review Fig. 9.21.

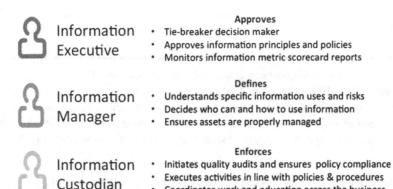

FIG. 9.21

Sample role descriptions

It shows a very simple representation of a DG framework in which the word stewardship is not mentioned. Fig. 9.22 illustrates how one organization connected the concept to its DG framework. Note that not only is there a clear distinction of accountability, but there is also a universal concept of stewardship. The concept of stewardship may or may not be one of accountability and responsibility. Stewardship is not a narrow definition and should be adjusted as your organization sees fit.

---

[7]Kudos to those people I have worked with over the years that, to a person, have all had to do double duty. There are many hard-working people in data and information management, and I have never seen a management team allow the designated DG deployment team to offload their current duties. Of course, it drags things out, but they hang in there. As for the leadership who demands the double duty (and does not offer additional incentive) while at the same time saying how important DG is, and often choosing to not engage adequately, stop being unrealistic and start to look at the human element.

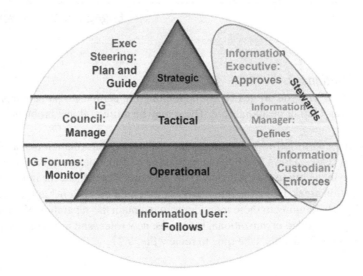

FIG. 9.22

Universal stewardship

At this point there will be questions about "who are the stewards?"

Stewardship is multidimensional. A common error is to declare an individual as the steward of a subject or content area (e.g., "Bob in marketing is now the czar for customer data"). However, stewardship is *not* an individual role in this context. The context and manner in which data is used will make a huge difference in the required style and intensity of stewardship. Fig. 9.23 illustrates

Stewardship is a function of context, business model, and relevance of content area

		Data Usage Category	
		Data as an Event or Transaction, e.g. enroll a new Member, sell to a new Customer, or repair machinery	Data as a Domain, i.e. use information ABOUT the domain
Sample Data Context	External Visibility e.g. Compliance to regulators	Data is gathered legally	Data is accessed legally	A
	Cross Functional Visibility - Use across multiple departments and functions	Transaction must be accurate	Data used by many areas needs to be accurate	B
	Departmental Visibility - Use to accomplish localized goals	Data needs to be useful for departmental use	Data with departmental visibility needs to be blocked from other areas	C

FIG. 9.23

Types of stewardship

how a single subject—in this case—could easily have two or more parties officially accountable for some aspect of customer data. Row A shows customers in an external context. Rows B and C show two different internal views of customer data. We based this figure on a client example where a single subject area had not only multiple stewards, but also required a *customer steward-ship committee.*

Accountability needs to stand out. Stewardship can be considered a function, so in essence anyone who uses or touches data in any way can be viewed as a steward. However, this certainly misses the whole accountability concept. Unless you want to solely state that stewardship equals accountability, you are going to need to call out who is accountable. In my experience, the best DG frameworks declare everyone stewards and then have separate titles for layers of account-ability and responsibility. This is one of those items that may not line up with other processes for deploying DG.

This activity will often occur in parallel with the presentation and approval of the DG framework. In very large and politically charged organizations, you will most likely identify personnel "as you go." That is, you will identify the DG personnel required for a particular portion of the information assets, such as the MDM project or AI project. This activity will be revisited often as DG expands or personnel change.

Ramification and benefits

Larger efforts will force the resources allocation issue at this point. Even though this is the first attempt at naming the roles, it could be controversial. This is why low-profile efforts are much more popular. You are entering the realm of making changes, and a larger profile effort has no choice but to confront these challenges. The low-profile effort will use the staff that are assigned to the initial use cases. At some point however, noninvasive WILL BECOME invasive. DG needs to go enterprise wide, or else you just did a few projects with a new approach.

This activity prepares the organization for the initial "bump" or learning curve that gets DG started. You will now know who is affected in terms of new job responsibilities. You also will get another in-dication of how serious the organization is, simply because the individuals who will be the best in the DG roles are most likely in high demand.

Some ramifications will appear that may be new to the DG team. For example—HR gives the DG team a blank stare. Often, taking the new responsibilities documentation to HR areas means discover-ing it has been a long time since new roles have been presented from another business area. While HR staff appreciates the need to manage people and have useful job descriptions, they do not do it very often, and they do not understand DG.

1. *Boundary problems*—The DG team may be accused of overstepping its charter by recommending organizational actions. This is something that can be avoided by early and frequent managing of expectations.
2. *Political considerations*—Inevitably, some areas of an organization will have more power and influence than others. The DG team will need to figure out the political situation and either work with the more powerful areas as allies or get executive assistance to counter any pushback. Politics will find you if you do not watch out for it.

3. *Incentives*—Corporate incentives are often used to move organizations toward better DG. HR will need to approve these and may even be useful in identifying programs that can provide a stimulus to DG acceptance.

Sample output

Below is a nice representation of a role description (one of possibly many) for Rocky Health. This is the DG chairperson.

- Leads discussion within the DG reference framework of standards, processes, and precedents
- Forms and leads a collaborative agenda balancing all parties priorities
- Recommends and referees ways forward on issues and escalation
- Assures required artifacts and repositories are kept up-to-date and/or statuses are reported (by the Data Governance Coordinator)
- Shows leadership in developing the DG program
- Coordinates gathering and review of DG metrics at the enterprise level
- Coordinates and monitors resolution of DG issues and project activities related to core data
- Monitors DG implementation roadmap/milestones and works with DG Leads to design and implement future roadmap items
- Monitors metadata health for core data
- Best Practice is to manage this role as a "rotating" chair

Two useful hints stand out for this activity:

1. *Don't be afraid of some horse-trading.* That is, if you want a certain individual to be a steward, look for opportunities to provide backfill for them. If a politically powerful area wants to dominate the councils, then request they become full-bore sponsors and take on accountability.
2. *Now is the time to consider some incentives.* If accountability means holding a manager to particular data quality targets, then work with HR to tie the DQ metrics into their compensation.

Some of the obstacles you may encounter and the useful responses to them are:

- *Perceived political threats from some getting "power" over data*—show how everyone is subject to DG, not just particular areas.
- *Human capital (or HR) concerns on changing job descriptions*—convene a meeting between your executive sponsor and the head of HR. This is a core business issue requiring executive input.
- *Fear that adding additional responsibilities will damage current productivity*—there is a learning curve, so offering to backfill will help. Also, reinforce that the "extra time" is not permanent.

The one singular hint that the reader should take away is *never* accept a DG resource that is someone who is willingly offered and is known to be a poor performer. Do not accept stewards who are people that cannot get anything else done. Even if they would make good stewards, they will not have the required respect from their peers. You do not want the DG program to be a dumping ground for unwanted staff.

The last important tip is to not shortchange what it takes to be in a custodial or stewardship position. Someone in a management position may be qualified to manage a functional area, but totally lack the awareness and insight to tackle DG. Essential skills for anyone embracing a stewardship role are:

- Understanding of organization needs and culture
- Commitment to organization success
- Desire to learn how to improve their organization

Socialize operating framework, engagement models, and workflow

You are now at a critical step in the process to activate DG. At a minimum, you need to review a summarized version of your operating model, functions, and any engagement models. The DG functions are new activities that will seem disruptive to many stakeholders (even though they are not). Also, it is time to present the list of proposed players and functionality. Review and approval of the DG framework and participants is more than one presentation to management. Typically, there will be some back and forth regarding roles and availability of higher-level resources.

The realization of new accountabilities will often stall DG efforts at this point, even if the full "green light" was given. Therefore, this is not a casual presentation. While you are looking for acceptance in principle and understanding of the details, you are not looking for a detailed review of all processes. You will need to explain roles, impacts, and "days in the life (again)." You will not get acceptance of all of the stewards and other ideal personnel you want. So, there will be some back and forth, and there will be gaps between presentation and approval.

Some aspects will be approved, and you can proceed with any roadmap and implementation activity. But there may be gaps in resources, so the roadmap or iterations of low profile efforts will need to work around that.

For a large program the DG team needs to be aware that this step may take time, as the approval of such structures usually occurs among personnel who only get together once per month, at best.

During the rapid rise of data science as a position within organizations, I noticed another phenomena popping up. That is a tone of indifference to DG personnel by highly educated data scientists. It has bordered on arrogance at times. I am not singling out data scientists, however. What is happening is more and more personnel are becoming data literate, and in lieu of formal guidance on how to behave, they develop their own view of a data driven world. Watch out for this. Therefore the data ethics capability now exists where it did not in the first edition. DG may need to define behavior codes and standards for everyone.

Approach considerations

There is not any exotic approach here. Prepare culturally acceptable material and present it. There will be different types of presentations depending on audience.

Keep the reviewers focused on the accountabilities and responsibilities. Are they appropriate? Are there cultural or policy-level barriers to the potential management layers?

Try to brainstorm the various types of questions that may be presented to the DG team. Remember, most reviewers will go immediately into forming mental organization charts. It is very easy to pick the operating and engagement models apart if that is allowed to happen.

Be careful when presenting the results of this activity to leadership. Very often too much detail is presented.

Ramification and benefits

The obvious benefit in this activity is either increased buy-in or reinforcement of existing buy-in. You will also be able to see the reaction of management to the functional model for DG. There are a few benefits of demonstrating to management a visible picture of what needs to happen to rein in data.

Many of the functions you are defining already happen in multiple places and are redundant or create conflicts.

The ones that are not occurring open the organization to risk.

The functions that are occurring currently should also be made as efficient as possible.

Most of the time the responses range from "this is a lot," to "is this what we are supposed to do?" You will need to be very clear with leadership that these activities are, more or less, *already* done. And they will be deployed incrementally.

Obviously, this is where low profile approaches have a great advantage. You rarely get into this type of conversation. But there are many examples, driven by regulation or a burning platform, where organizations need to look at and address the magnitude of becoming data driven and managing data assets.

Sample output

See the examples in the case studies below.

Architecture and design case study—Rocky Health

Given the desire to expand DG the new operating framework and engagement models are important to the effort. Tom needs to make sure the use cases can be addressed while creating a sustainable operating model. Architecture and Design must not only address use case DG but also long-term operations of a larger program.

Capabilities will come from two places—the use cases, then whatever else might be needed for longer term sustainability. There are three use cases: Patient Access, Outcomes Metrics, and Financial Performance. Tom will need to prioritize them so the roadmap can have realistic timing, but for now the operating model for DG needs to consider all of the capabilities for these use cases. In addition, some other capabilities may be needed.

Use Case	Capability
Patient Access	Accountability (Stewardship)
	Metrics standardization
	Data standardization
	Reference data
Outcomes Metrics	Metrics standardization
	Reference data
	Data quality

(Continued)

Use Case	Capability
Financial Performance	Data access Accountability (Stewardship) Metrics standardization

The preliminary capabilities (from alignment activities in Strategy) quickly expand once the team starts to examine the use cases. (This is pretty common—once you step away from low profile things expand quickly.) You can see how a picture is developing of what needs to be done to manage data and enable business goals. Obviously, some sort of metrics glossary is required. Also reference data.

These capabilities align with business needs, and once the use case is prioritized, you have prioritized the sequence of capabilities. Then the operating model activities can expand these capabilities into functions, or processes.

Tool requirements appear as well. Tom needs to find out if Rocky Health has some sort of supporting technology for metrics, data standards, or reference data. He can plan to define requirements for tools, and adopt any internal technology or plan for new technology.

Lastly, Tom needs to design an operating model that can expand and sustain the required data practices required by the use cases. Whereas the initial operating model was Tom overseeing the data issue process, multiple use cases will require a mechanism to balance issues vs new capabilities and support the DG program when resistance or other change issues are encountered.

The Rocky Health team, like many others in this situation, had an intense discussion on whether an MSOM is required.

Low profile or noninvasive efforts may already have achieved a minimal sustainable status. Then again, if they are more proof of concept, then you may need to still look for an initial operating model that represents a minimal sustainable state. Tom decided that another operating model representing a minimal state would be needed, as well as the longer-term operating model (Fig. 9.24).

Architecture and design case study—Rocky regional electric coop

Diana, although having a history and attempt at DM, is really in new territory. There are a lot of efforts that require data management and governance capabilities. The challenge with Rocky energy is not so much what capabilities are needed. A quick review of the use cases will prove they need almost everything off of the capabilities list. The deeper issue is how to keep DG moving and demonstrating value, vs DG being sucked into the projects. Tying DG to projects is a powerful method to implement DG, but extra care is required to prevent DG from becoming a project delivery resource.

The use cases could become major projects in their own right.

1. Enterprise reporting and business intelligence—The broad span of the effort could require several new and significant data policies
2. Engineering and asset management—The use of an industry standard model is good, but DG could make it great.
3. Customer service hook up—Any operating model will need to accommodate the operational aspects of DM such as MDM and data quality.

Current State

Executive Sponsor

Data Governance Team

Data Issue Working Group

New MSOM

Executing Sponsor

Data Governance Council

Data Working Group   Data Working Group

Long-term DG Operating model

Executive DG Oversight

Data Governance Council

Data Working Group(s)

Data Working Group | Data Working Group

IT | PMO | Others

Data Stakeholders

Organization	Membership and Responsibilities	Accountabilities
Executive Sponsor (Tom)	• Provide strategic direction • Resolve escalated issues	• Ensure resources and priority
Data Governance Team	• Carry out initial DG activities	• Communicate and implement data decisions
Data Issue Working Group	• Participate in data issue tracking	• Determine best possible resolution to specific data issue
Executive Sponsor (New sponsor)	• Provide strategic direction • Resolve escalated issues	• Ensure resources and priority
Data Governance Council (DGC)	• Make data decisions based on input from Working Groups and DGO • Chair and/or participate in Working Groups as appropriate	• Communicate and implement data decisions • Determine best possible resolution to specific data issue
Data Working Group(s) -Ad Hoc	• Chaired by a DGC member • Focused on Data Issue or Data Projects or data use • Includes Subjects Matter Experts from any level of the organization	
Executive DG Oversight	• Provide strategic direction • Resolve escalated issues • Ensure adequate resource availability	• Ensure resources and priority
Data Governance Council (DGC)	• Make data decisions based on input from Working Groups and DGO • Chair and/or participate in Working Groups as appropriate	• Communicate and implement data decisions
Data Working Group(s) -Ad Hoc	• Chaired by a DGC member • Focused on a data issue or data project • Includes Subject Matter Experts from any level of the organization • Representatives of reporting and analytics organizations • Identify duplicative and conflicting data requests • Consolidate intake request processes	• Determine best possible resolution to specific data issue • Provide cross-functional oversight to data projects
Data Stakeholders	• Community of LOB people with specific day-to-day accountabilities for specific data • Attend / participate in DG updates • Be a positive influence for DG in the organization	• Data quality and accuracy as data is created

FIG. 9.24

Rocky health sample operating model

The operating model required some serious consideration. Diana and her team realized they need a many-iteration roll-out of DG. Also, an MSOM will need to be enterprise-wide, but simple to manage and no overhead implied. It was determined the MSOM would be an oversight body for all data projects. Diana sees the opportunity for embedding data management and governance in IT project as soon as possible. Helping the CIO will also serve to start to eliminate the bad feelings from the first attempt at DG.

The Engineering and Asset management effort was selected as the first visible use case. The operating MSOM was directed at that use case (Fig. 9.17 is the RREC Operating model, and the engagement models with IT and development are Fig. 9.19).

Summary

> *Every great architect is - necessarily - a great poet. He must be a great original interpreter of his time, his day, his age.*
> **Frank Lloyd Wright**

At some point, DG needs to make the transition from a cool concept to an operational model. This section offers the tasks to accomplish the first half of that exercise. It is the step where the realities of DG start to run into the realities of organizations realizing the need to approach information differently.

As a result, some engineering and architecture is required. Relevant principles of data and information management are applied. The principles form a foundation for belief that is critical. They also provide implications and rationale, which then frames policy development. Then the required capabilities are turned into functions for plan, design, management, and operation of DG. From the principles to the detailed processes, you need to leave this activity with a clear understanding of an operational model and presentation of DG for your enterprise.

Essential questions

1. How many layers should there be in an operating model?
2. Why is capability modeling so useful for DG design?
3. Do you always need a minimum sustainable operating model? When do you and when do you not need one?
4. Explain why DG is a business capability.

Implementation

10

Chapter Outline

> *Everything starts somewhere, though many physicists disagree. But people have always been dimly aware of the problem with the start of things. They wonder how the snowplough driver gets to work, or how the makers of dictionaries look up the spelling of words.*
> **Terry Pratchett, Hogfather**

Overview

The Implementation work area is where all of the "get started" activities live. The implementation activities, therefore, must not only produce the list of events required to deploy data governance (DG), it must also provide an outline for success and sustainability.

Almost any organization using automation for information processing has tried to do something formal to manage that information. Remember, management of information already happens in organizations, but it just happens poorly. DG fixes that.

Data Governance. https://doi.org/10.1016/B978-0-12-815831-9.00010-2
© 2020 Elsevier Inc. All rights reserved.

The "roadmap" for DG is the penultimate work product for the DG program. That is, aside from the operating program itself, it is the most popular output from the DG team's deployment efforts. It also forms the foundation for the sustainability of DG.

Sustainability means acting to ensure that the right processes are in place by which the DG organizational framework will *continue* to perform the governance function. Core to this requirement is the overlooked fact that the organization *accepts* the governance of data—that the function be managed, its results be monitored and measured, and the obstacles that so often cause DG programs to falter or fail are overcome. Very few organizations think ahead about what needs to happen 1 or 2 years down the road (Fig. 10.1).

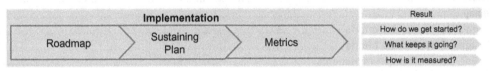

FIG. 10.1

Implementation work area

Measuring change adoption and managing the required behavior changes are only a few pieces of the puzzle. Other critical components include developing and measuring the metrics that reinforce DG's value and track progress, having clear principles and policies documented and in place, and verifying that the organization has the resources required to support DG after it is rolled out.

If you don't have one already, you will need to put in place a formal *organizational change management* (OCM) program to sustain DG in your organization. Several OCM activities are described in this chapter but try and start them ASAP. Put them into your checklists as soon as your approach permits.

> Ideally, you should start planning for changes as soon as you know what you are planning to change. For example, if you identify new capabilities as a requirement or as part of your vision, you need to start to consider if your organization is ready for that change. Hence the change capacity assessment under the Strategy work area. If you have not socialized any of the new things that are planned, and you are getting ready to implement a program, you have a problem. The change management activities in this work area are to implement additional actions to ensure sustainability.

The formality and discipline inherent in DG is new and different for many organizations—and difference means *change*. Change requires that people adjust their behaviors to the new way of doing things, and changing behavior is no easy task—just ask those of us who make (and break) those New Year's resolutions every year! It won't happen just because you say it will or believe it's the right thing to do. People naturally resist change because they are afraid of it: afraid it will be hard, or they will fail in the new world, or lose something—power, competence, or influence—to name a few. You will have to overcome that resistance in order for DG to be successful and adopted by the organization. That formal OCM program, with the right executive sponsor, is critical to helping you accomplish that.

Organizational change management is a well-known discipline within the realm of *organizational effectiveness*. Most OCM work is based on the Plan, Do, Act model.

This activity area contains the Planning aspects of OCM. These can be selected for your checklist and done as early as possible in your overall effort. I put them here because this is the last possible opportunity to consider OCM. Once you start to implement and operate it is too late.

Fig. 10.2 show the types of activity in this area.

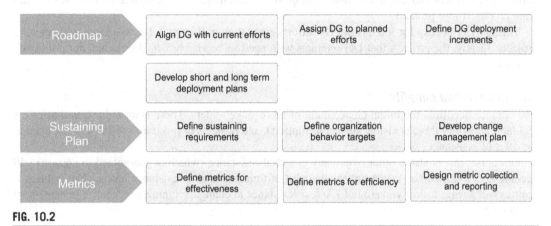

FIG. 10.2

Types of activity in Implementation

Roadmap

The most frequently requested deliverable in my practice, bar none, is a roadmap. This is not hard to understand. "Where do I start?" is a legitimate and important question. Defining and presenting a roadmap can be 80% engineering and 20% art. You need to consider not only what DG steps are required, but the interaction with all other activity that any normal organization has in play. There will be many contexts within a roadmap—people, processes, operations, data, architecture, methods, etc. All need to be considered. You also need to consider reasonable short and long horizons. A lot of organizations ask for a 5-year plan. I often ask them "Have you ever made it to the 5th year of any plan?"

Align DG with current efforts

Remember that you need to govern something, so this activity starts the process to see what specific projects and programs DG will support and oversee. That means taking a close look at what is happening currently with data intensive projects. There is a temptation to "grandfather" or exclude all work in process, providing an exemption from DG scrutiny. However, you might miss an opportunity to really add value, and assist a project that is struggling. In addition, if projects such as analytics or master data management (MDM) are underway without DG, then you can literally prevent them failing. A large enterprise resource planning (ERP) effort may require many types of DG oversight.

Caution is in order. The worst thing that can happen is to overwhelm a deadline-driven effort with activity that seems to be interfering. The most common areas to resist DG are always an applications or project-delivery area. Therefore, the DG team needs to choose activity where DG can assist the project in achieving its goals, and not appear to be interfering. This may seem like "giving up" on enterprise DG, but it is not. Rather, this is a practical implementation strategy.

Approach considerations

The most important aspect of this activity is, if you need to do it, you also will most likely need to do the next activity listed below—that is, you look to an align DG as a possible means to assist current projects. They are separated in the event you are tackling something pretty large, and you need to make a list of projects that DG *will not be able to help*. If DG can help, then the next activity serves to integrate DG into those projects.

This activity is easy for low-profile DG efforts since they will already be associated with specific projects. Ideally, the projects that DG interacts with need to be visible but, if possible, not politically charged.

Ramification and benefits

The obvious benefit is the initial engagement of more stakeholders. There is an indication that DG is real (and hopefully there is visible sufficient support), and projects or programs will at least be open to the offer of assistance.

Of course, this is also a point where any potential resistance (which is guaranteed to happen) will become very evident. You may want to take a bit of time and assess resistance levels. (Not politics—politics will always follow you around.) Assess resistance to change. Organization change management has started.

Sample output

See Fig. 10.3.

Assign DG to planned efforts

The integration with other efforts means getting embedded into other project plans, not observing, or suggesting. Your DG team will need to sit down and wade in with the "lucky candidates." Likely outcomes will be educating projects stakeholders subject to standards and governance, then refine the project governance bodies and committees. There may be some sort of tweaking of stewardship-like roles as well.

Whatever the type of project or interaction you determine at this point, build it *into* project plans. Glance at Fig. 10.4 and notice that projects and DG are closely aligned. Rather than hazy Gantt charts where you guess and approximate, set some measurable objectives.

Always emphasize that DG is there to help. It will not hurt to remind stakeholders you are looking for the "net zero" or minimal impact. This is not anything new—it is different.

Approach considerations

A project where careers are at stake may not be open to external governance. However, a good dose of DG can help ensure success. When DG is built into high-visibility efforts, it is a matter of leverage and coordination as opposed to additional oversight. One goal of low profile approaches is to prove there is no threat. Then the larger project can embrace DG.

If you have a data centric program, like advanced analytics, artificial intelligence (AI), MDM, etc. program and DG is being initiated in the context of a formal data management (DM) office, make sure you keep DM leadership informed, and adjust the various DG area charters to ensure the selected project and programs can fit into the charter for DG.

RREC Project Plan Realignment

Project	Current Plan(s)	Data Managed Roadmap tasks
Enterprise reporting and business intelligence (BI)		
	Assess tool needs	Match BI and Reporting to business needs
	Report inventory	Report inventory
	Training	Integrate BI and Reporting with governed self-service
		Roll out common reference data
Engineering and asset management		
	Create engineering standards	Create data standards for equipment and data
	Define asset management policies	Define stewards for asset data, and other subject areas
	Define new network design and management procedures	Define asset and network data collaboration processes
Customer service hook up		
	Clean up customer data	Define Customer MDM plan
	SWAT team for new connection	Profile customer and related data
	Hire customer data quality control staff	Implement Customer Data Quality program
		Change internal workflow to avoid hiring

FIG. 10.3

RREC project alignment

Each project or program selected will require identification of the accountable and responsible parties for DG (i.e., the stewards and custodians). They should get a good review session of why they were picked and start to make sure they are amenable to the training and communications plans.

Large ERP efforts frequently have a DG program built in by the systems integrator that has been hired. In fact, your best friend may be the integrator who is overseeing the large program or project. Most large integrators (e.g., Accenture, Deloitte, etc.) bring DG, in some form, to every program they run. But exercise some due diligence and make sure their DG is truly DG. Since the first edition I have

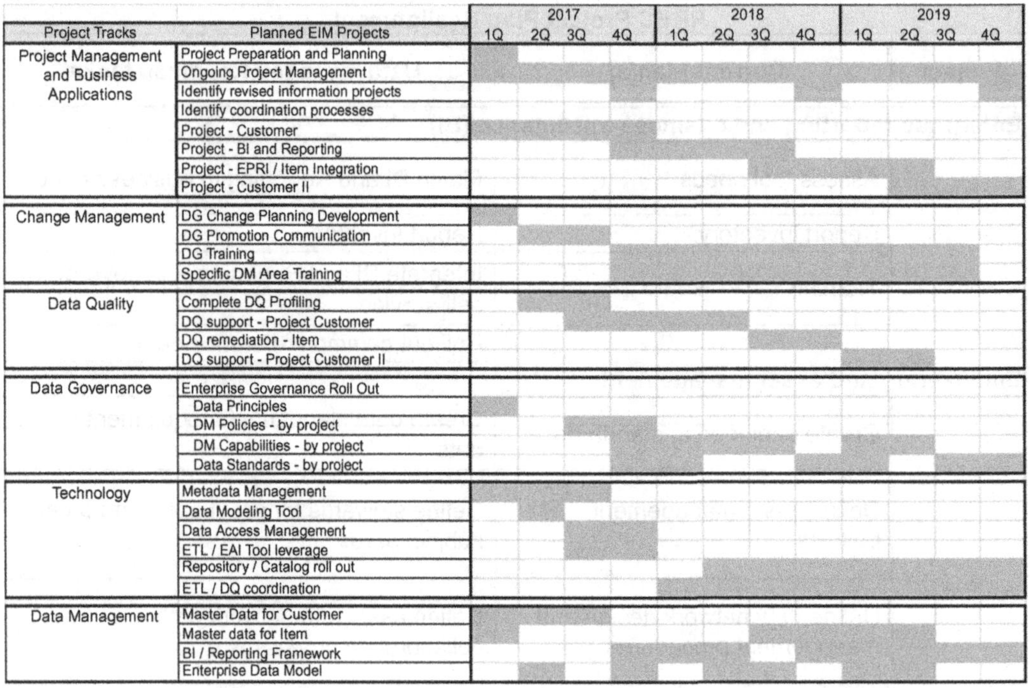

Project Tracks	Planned EIM Projects	2017 1Q	2Q	3Q	4Q	2018 1Q	2Q	3Q	4Q	2019 1Q	2Q	3Q	4Q
Project Management and Business Applications	Project Preparation and Planning												
	Ongoing Project Management												
	Identify revised information projects												
	Identify coordination processes												
	Project - Customer												
	Project - BI and Reporting												
	Project - EPRI / Item Integration												
	Project - Customer II												
Change Management	DG Change Planning Development												
	DG Promotion Communication												
	DG Training												
	Specific DM Area Training												
Data Quality	Complete DQ Profiling												
	DQ support - Project Customer												
	DQ remediation - Item												
	DQ support - Project Customer II												
Data Governance	Enterprise Governance Roll Out												
	Data Principles												
	DM Policies - by project												
	DM Capabilities - by project												
	Data Standards - by project												
Technology	Metadata Management												
	Data Modeling Tool												
	Data Access Management												
	ETL / EAI Tool leverage												
	Repository / Catalog roll out												
	ETL / DQ coordination												
Data Management	Master Data for Customer												
	Master data for Item												
	BI / Reporting Framework												
	Enterprise Data Model												

FIG. 10.4

RREC roadmap

seen some DG services presented as add-on (read more $$$) services, for efforts where it should be mandatory and included.

Lastly, make sure the roadmap reflects this by showing the affected projects as well as the various DG efforts.

Sample output
See Fig. 10.4 for a sample of how a program interacts with DG. This view of the RREC case shows a typical road map that intertwines governed projects with the DG roll-out.

Ramification and benefits
The obvious benefit is the engagement of more stakeholders. There is an indication that DG is real (and hopefully there is visible sufficient support), and projects or programs will at least be open to the offer of assistance.

Again, perhaps to a greater extent, resistance will appear and your OCM plans need to be ready.

Define DG deployment increments

For all but the smallest, noninvasive effort, you will need to do some sort of definition of a roll-out plan. It is a core part of the roadmap. The important feature is the incremental approach and defining suitable increments. Obviously, you should not announce to everyone, "DG is here!" and then wait for DG to happen. Very often, it is too easy to develop a roll-out strategy that does just that. No matter what good intentions are held, the roll-out process for DG can very easily be perceived as force-feeding the organization. Remember, the "something" to be governed is perfectly fine to implement in small pieces, by project, or by program. Start an effort that does not ruffle feathers, get your OCM approach in order, then expand.

Most of the time, you will merge this activity and the next one (short- and long-term deployment plan) together to produce most of the roadmap. (Fig. 10.4 presents a nice overview of how this all comes together.)

If you are starting a second time (or third) and started as a large, monolithic program the first time, then finding smaller pieces and even some low profile, or noninvasive pieces, is probably the required approach.

This does not mean you have *separate* DG efforts for every project. You will need to plan and coordinate among projects to ensure the enterprise aspect of DG.

Approach considerations

Defining roadmap increments is a rationalization process. There will be a lot of input to consider. Obviously, if you are low profile, you won't be doing this. But eventually you will need to consider broadening the DG program. This means before you segment possible work, make sure you are looking at all the various efforts to see what can be combined, leveraged, or even isolated (to allow for short term wins) (see Fig. 10.6).

Once you have a collection of segments that can be rolled out, line up the DG requirements. Make sure you know what DG capabilities, process, and/or functions are required for each increment. Obviously, the schedule can follow the sorted-out increments.

This activity may also provide some feedback for possible revisions to the charter, roles, responsibilities, and operating models.

Ramification and benefits

Success and sustainability for DG will require early visibility. So, make sure the initial increments generate value and good impressions. This accomplishes two things:

1. The DG participants gain valuable experience and the organization is exposed to DG.
2. The early efforts in governing result in feedback that will indicate any required adjustments to training, staff, or operating model of DG.

Take a deliverables-based approach to this task. Even though you are rolling out a sustainable program, you still want to have the team work toward discrete work products and develop artifacts. It is easier for personnel new to DG to work toward a specific product. They will assimilate their knowledge by doing these and will mature more rapidly toward DG concepts.

Since this is the first set of tasks where the DG deployment group, the stewards, and the governed projects intersect, it may be useful to have some checkpoint meetings established. This is a good way to capture feedback and detect resistance without allowing it to ferment.

Sample output

Fig. 10.5 shows a sample of a coordinated series of short-term activities in conjunction with larger projects and the overall roll-out of DG. Please note, some of the activities in this figure reflect change management tasks as well as targeted governance activity.

Farfel Data Governance Milestones & Roll Out

January-12	February-12	March-12	April-12
Stand Up Corporate Stewardship			
Initial training - Data Governance Council (DGC)	Report of Capacity to adapt all aspects of IM and DG	Deployed DG /IM Roll Out	DG/IM/ EA Gap and Framework Direction
Initial Training - Custodians and Stewards	Preliminary Organizational Change Management Requirements	Refinements to DG /IM Organization	Deployed DG /IM Roll Out
Stakeholder Assessment	DG Operational Training Requirements - How to act; new process & behavior change	DG Sustaining Checkpoint	Refinements to DG /IM Organization
Communications Requirements	Approved DG Principles and Policies	Final DG /IM Stand up Measurement and Feedback plan	DGC Meeting
Training Requirements	DG Organization official kickoff	Enterprise reference data approach	Custodian / Steward Meeting
Approved Functional DG/IM Model & Charters	Final Change Management Requirements	Define modifications to SDLC /AGILE methods reflecting DG	Enterprise reference data approach
Finalized list corporate stewards and custodians	Final DG /IM Model & Charters	Enterprise metrics catalog reviewed	Enterprise metrics catalog reviewed
Rationalized Farfel DG Principles and Policies	Final DG /IM Stand up Communications Plan	Modifications to SDLC /AGILE methods reflecting DG	Execute DG Organization Change Management
Custodian / Steward Initial tasks *	Final DG /IM Stand up Training Plan	Enterprise data model - enhanced	DG Tools and Process roll out
Rationalized Farfel DG Principles and Policies	Final DG /IM Resistance Management Plan		
First group orientation for stewards and custodians			
DGC Initial tasks	**DGC Initial tasks**		
Address initial issues to test elevation process	Review ERP / DG Integration RACI		
ERP Project interface consistency	Review ERP Issue log		
Item / Inventory definition issues			
SharePoint management	Steward Custodian Initial tasks		
Review IM and DG functions (the "V")	Attend Stewardship and Custodian Education		
Attend Stewardship and Custodian Education (walk the talk)	Address initial issues to test elevation process		
	Data Warehouse enhancement		
	Use of enterprise data model		
ERP Program Support			
DG Integration	ERP DG RACI Deployed	ERP data flow / interface support	ERP / DG Organization Change Integration Plan
ERP DG interaction RACI	ERP portion enterprise data model	ERP metrics and BIR catalog	ERP / DGC Leadership Coordination meeting
ERP / DG Stakeholder Analysis	Verified list of ERP Stewards and Custodians	ERP Function and Service support (FD)	
ERP Subject Area Definitions and DG Scope	ERP data flow / interface support	ERP DQ standards	
Laser Portion Enterprise Model	ERP DQ standards	ERP MDM Standards	
ERP metrics and BIR catalog	ERP MDM Standards		
ERP DQ standards			
ERP MDM Standards			

FIG. 10.5

Coordinating data governance activity

Develop short- and long-term deployment plans

Even after the roadmap is set up with increments and sequencing, you may need to focus on short-term or long-term detailed plans. This task is optional for most approaches to DG. But occasionally, really detailed planning is necessary. Budget-sensitive efforts might require more details, for example, see Fig. 10.6.

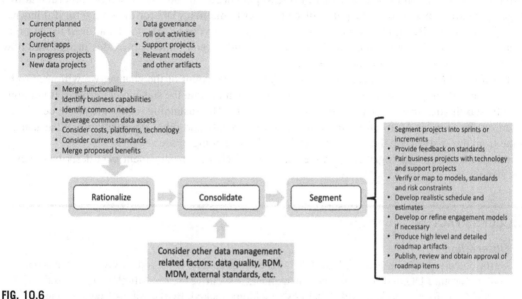

FIG. 10.6

Roadmap rationalization

Approach considerations

One key aspect of a roll-out strategy is to get the governance structure busy doing some kind of DG activity. Too often, the stakeholders are sent to training and then told to wait for the first opportunity to govern. This activity, therefore, not only refines the incremental events that deploy DG, but also specifies what is happening in these events, as well as fine-tuning the operating model to match the roll-out approach.

Ramification and benefits

The short-term plans from this task will most likely be old-fashioned project plans; joining the stand-up of DG with value-added projects.

Sample output

There are sample detailed roadmaps in the appendices.

Sustaining plan

> *Nothing travels faster than the speed of light, with the possible exception of bad news, which obeys*
> *its own special laws.*
> **Douglas Adams, Mostly Harmless**

Define sustaining requirements

At this point in the development of the DG road map, the team must have some idea of the extent of changes required to implement and sustain DG, especially if the change capacity and information management maturity assessments are being used. Socialization during Strategy and Architecture activities will also point out the area that the OCM plan needs to address. It will be important at this stage to identify what needs to be put in place in order to maintain DG for the long term if it has not been done already. Being able to maintain DG requires that your OCM plan be attuned to the organization's culture. The work done to identify the sustaining requirements provides insight into how those cultural elements need to be addressed.

For broader DG efforts, where there is some idea of the extent of the change, we will need to be sure we have developed a specific analysis of the change impact on the stakeholder groups. If not done already, stakeholder analysis may need to be inserted here. The sustaining requirements are developed based on the analysis of all affected parties (the assessments and other observations). What usually come from the analysis are the following needs for the sustaining phase.

All of these elements together form the basis for the change management plan described later in this chapter.

SUCCESS FACTOR

Observation
Unfortunately, OCM is a discipline that is rarely effectively deployed in organizations, particularly around DG and information management. Executives tend to think of it as "squishy" or "soft"; far from it, as the hard dollar costs (seldom tracked, by the way) of poorly managed change in an organization are significant. If you want to realize any benefits from your DG implementation, you MUST have an OCM program to drive the required behavior changes. Otherwise, you are wasting your investment in DG now, because it won't stick, *period*. Think about it—is this really the first time your organization has tried DG? It's a "pay me now or pay me later" scenario.

If you are on the second or third attempt, it is almost a sure bet that you did not have adequate OCM plans in place the first time around.

Approach considerations

There are many change management processes available to use (Prosci, Kotter, Bridges, etc.). All contain basically the same elements, and all are effective when deployed properly. Using an accepted, published process also allows you to insert these activities into your approach to DG sooner. Very often some of the OCM activities start in Strategy. When you get to the Implementation point in your effort, the requirements for sustaining the program become evident. Your sustaining requirements should be in a context that:

- Focuses on engagement and managing resistance
- Follows best practices and provides metrics for consideration
- Offers sample tools for planning, assessment, and support of stakeholders and sponsors

Remember most approaches to change management are based on the Plan, Do, Act model.

1. *Planning*—assessing the need for change and developing the approach and detailed plan to manage change. Planning needs to be finalized as early as possible, so while the checklist activities are reviewed here, the sooner you can do them the better.
2. *Doing*—executing the OCM plan to help people transition from the "old" state of work to the new. Doing change means rolling out the plan. Executing communications and training events for example. Aligning leadership and analyzing stakeholders.
3. *Act or Sustaining*—implementing the mechanisms and structures to ensure there is no reversion back to old ways! It is easy to confuse DO and ACT. On-going communication and training is certainly a part of change management. But you also need to monitor the actual changed activities and look for effectiveness of new behaviors.

Picking the right sponsor for DG (i.e., from the business) is essential. Per the Prosci Best Practices surveys since 2003, the right sponsor has been the number one success factor for any change effort. That person has the influence and political capital to make things happen; get him or her engaged very early on. The right sponsor is an essential OCM "best practice" and must be addressed early on. Without a good sponsor your chances of success are slim. Also, in most organizations IT usually does not have the credibility to sponsor something like DG. Go after a business executive and keep pushing until you get the right one.

- Other items to consider when determine the OCM requirements are:
- Are there any other assessments you can use? Many organizations do frequent employee surveys.
- Did your change capacity assessment provide enough insight? If not, leadership alignment and stakeholder analysis can be used to beef up your discovery. Some steps you might need to take are:

How do you need to staff OCM? Visible efforts, like RREC case study, will require some change agents and other resources. Low profile efforts may be ok with only the existing stakeholders. Larger efforts will need to treat obtaining a sponsor in the same way as hiring someone—qualifications and experience.

Stakeholder analysis should be done considering all those who are impacted, to what degree, and what their likely reaction(s) will be. It will be important to understand how people will react so you can develop methods or approaches to address their resistance or engage their support.

Make sure you spend adequate time on communications requirements. Open, honest, and frequent communication is absolutely critical—and it is not a list of required PowerPoint slides. Various stakeholder groups will require different and differing levels of communication, and the opportunity to provide feedback. Communication *must be two-way*. Only if you know what people are thinking or how they are reacting will you be able to "course correct" your plans and address the issues.

What kind of resistance will there be? Proactively identify specific types of resistance (overt, passive, etc.) and identify the required activity to deal with it. Planning to manage resistance is essential. It is out there and cannot be ignored or it will undermine all your efforts—guaranteed. There will be varying levels of resistance, from openly hostile to passive (Fig. 10.7).

The important thing is to understand why people are resisting and to try to address it. Considerable change requirements can result from understanding what types of resistance you will encounter (Fig. 10.8).

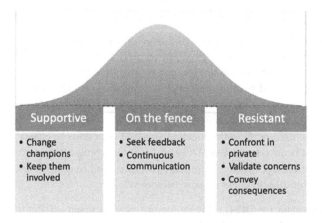

FIG. 10.7

Resistance spectrum

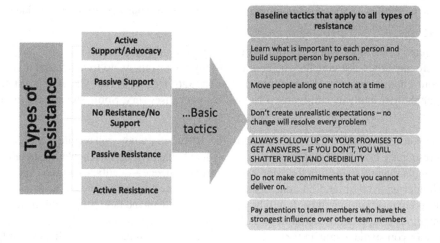

FIG. 10.8

Resistance types

Answering "What's in it for me?" (WIIFM) is an important OCM principle. It helps people connect to what is happening and move through their resistance to support. Determining WIIFM is a key requirement.

If you have been doing this for a while, you become accustomed to resistance in all its forms. However, as someone who may be new, please bear this in mind: many of the behaviors that DG deems risky (e.g., departmental databases with mission-critical spreadsheets copied to USB drives, etc.), are viewed as necessary and acceptable. Keep a positive outlook; identify available incentives and provide education on the benefits and rationale for doing DG. Engage people in the process as much as possible. Keep negative responses as a tool but deploy them only after trying the positive.

The DG team can benefit greatly from the help of an organizational development specialist if one is available. They can define the sustaining requirements and the change management plan. Many HR organizations have a unit of these professionals to help with change efforts. Make use of them if you have such a group in your organization.

Ramification and benefits

It has been said previously but is worth reinforcing once again—the *root cause* for failure of many DG or other related data-type programs is the failure to recognize that organizational changes *must* be proactively managed. If you do not manage the movement from the current state of organizational behavior around DG (scattered, inconsistent, or nonexistent) to the desired future state, you will fail. You need to manage the psychology of new behaviors.

Sample output

This is an outline of change requirements—examples of which are in the appendices:

- Communication plans
- Education and training plans (by type of stakeholder and degree of impact)
- Sponsorship expectations and guidelines
- Individual coaching plans for executives to enable them to effectively support required changes
- A resistance management plan and tactics
- Process and policy alignment plans
- Organizational realignment plans: structure, roles

The appendices contain examples and templates of the following outputs:

- Stakeholder Analysis Grid
- Change Capacity Assessment
- Leadership Alignment Assessment
- Metrics for Sustaining DG

Define organization behavior targets

I added this short section to help clarify what behavior changes are required. Processes and functions are one thing. How to do them is another. For larger efforts you will need to articulate new behaviors. Low profile DG will remain very specific about who does what until broader recognition of DG starts. Then you will also need to consider describing behaviors. For example, AI and advanced analytics enters an organization via a stand-alone effort. Then data monetization takes hold. Suddenly the CEO insists everyone be on the lookout for monetization opportunity and "manage data better." The problem is no one knows what that looks like. "Use the glossary" is an inadequate expression of what to do.

Approach considerations

Essentially, you are describing what data literacy looks like IN ACTION. First thing to consider is "does your audience actually understand the concepts that they are being asked to adopt?" Then you need to articulate how that understanding will appear within business-as-usual activity. How do you measure using a glossary? (Yes, this activity will suggest some possible metrics for DG use.) Is the training for glossary users going to cover the concepts behind actually why you need a glossary? Every capability will need to offer some solid descriptions.

Here are some simple steps to come up with a list of behaviors (by capability):

1. Measure if there is rote-level and understanding of DG.
2. Take necessary steps to achieve understanding—that is, stakeholders understand what DG is and how it will add value. For those of you tacking DG a second time (or more) this is really important—definitely add this to your checklists. Most likely you did not achieve this on the first attempt.
3. Describe common, specific work flow for interacting with DG—for example, how to use a glossary as a business person vs a data custodian/caretaker or developer.
4. Look for and assist your stakeholders in getting their behaviors around DM and governance to be instinctive and automatic.

Ramification and benefits

Increasing data literacy can do nothing but further increase the likelihood of success. In addition, you develop a solid constituency of personnel who can talk positively about DG.

Sample output

Below is a sampling of capabilities from Rocky Health and RREC. There are a lot more than presented, but this will give you the general idea. Obviously, you would do as many as required to develop understanding of desired behaviors:

Rocky Health sample behavior list		
		Desired VISIBLE behaviors
	Data governance capabilities	
Plan	Data and governance strategy	
	EIM and DG business alignment	Active data governance input for budgets and strategic planning
	EIM and DG goal setting	Approved annual DG/DM targets
	Data principles	Visible consideration of data principles within app design and development
Define	Data governance requirements and design	
	Compliance identification	Compliance proactively works with data governance
	Data governance frameworks	
	Operating framework definition	New operating framework has engaged new participants
	Roles and responsibilities	New roles are active and adding value
Manage	Data governance operation	
	Data access and user	Users are accessing correct data sources
	Security/privacy governance oversight	Data privacy is considered for all patient record use
	Literacy and maturity targets	DG measures data literacy

		Desired VISIBLE behaviors
Data management capabilities		
Define	Define EIM components	
	Metrics and KPI	Clear understanding and use of standardized KPIs and dashboards
Manage	Data movement and integration	BI understands and uses policies for data sourcing
Operate	Data development/test/production	AppDev needs to put data management into test plans

RREC sample behavior list		
Data governance capabilities		**Desired VISIBLE behaviors**
Plan	Data and governance strategy	
	EIM and DG business alignment	Active data governance input for budgets and strategic planning for the major initiatives
	EIM and DG goal setting	Approved annual DG/DM targets by initiative or any other targets
	Business strategy support	Ensure that department heads understand the role of data in business strategy
Define	Data governance requirements and design	
	Federation requirements	There will be periodic communications from new areas with new authorities
	Data governance frameworks	
	Engagement model definition	New work flow between app dev, BI, and various reporting areas will start to evolve. This needs to be reinforced and measured
	Data literacy	Literacy needs to follow the federated model, so stakeholders need to be aware of the need to learn new things
	Supporting technology	
	Metadata management	App Dev, BI and business analysts will need to interact with glossaries, model (EPRI), and other metadata structures. Training and measurement need to be explained as well as the actual hands-on interaction with metadata capabilities
	Data mastering	Should MDM result in a new Item Master, there will be significant effort to address the changes in work flow and processes
Manage	Data governance operation	
	Data standards	Standardization will be a big change for RREC. Stakeholders will need to understand how to incorporate new standards into their existing work

Continued

Data governance capabilities		Desired VISIBLE behaviors
Operate	Communication	
	Communicated expectations and accomplishments	
	Communicated data-related directives	An effort with the span of RREC will require extensive communications. Most of these will be to ensure that people understand new behaviors
	Data governance services	
	Data sharing agreement services	Large architectures and governance programs are trending to services. Services require different behavior in terms of support, use, and expectations

Develop change management plan

Earlier I mentioned that, while change management is presented here, it can start as soon as you need it. For organizations that are on their second attempt at DG, you need to seriously consider starting to address changes as soon as you have a change capacity survey done back in the Strategy activities. OR, during architecture and design as soon as you get a picture of how work will be different.

As soon as you have enough information to frame change then the detailed planning for managing the DG program changes can begin. A comprehensive plan and effective execution of the plan is the means by which you will build awareness, understanding, and acceptance of DG for the organization.

As you identify the tasks and timelines required to implement and sustain DG, remember you are defining a structured process, measures, and monitoring for integrating data. If your effort requires a change plan of any sort, then you need to apply the necessary rigor to make it a good change plan.

Approach considerations

When building your communication plan, think about bringing people up a "change curve" from basic awareness through understanding, to acceptance and commitment (see Fig. 10.9).

It takes time and repeated reinforcement to get people through this curve. Consider all of your audiences, the degree of impact to them, the key messages you want to deliver (and who should deliver them), sequencing, timing, and media. Generally, the "big picture" stuff should come from the executive level; changes to day-to-day process and procedure are best delivered from the direct management or supervisory level. Also, plan to collect feedback on how well your messages are getting through. Remember that communication is *two-way*; you have to listen as well as deliver information, and then demonstrate that feedback has been heard and acted upon.

Communication—always important, and rarely executed properly. Here is an example. A new policy is to "go live" on a certain date. Stakeholders are trained. Often too early. Even if done in a timely fashion, there is rarely any feedback. Then the Friday before the change goes in, an email blast goes out with a "reminder." This is poor communication.

If you have no idea or confirmation that anyone understands, is ready, or even supports what you are doing, you have not communicated. You need to witness a positive response that tells you communication has occurred. If you don't ask, you have no idea.

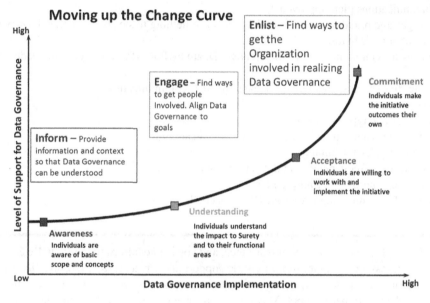

FIG. 10.9

Basic change curves

When rolling out a DG program, it is essential to be specific about what is changing so that those impacted understand precisely what to start, to stop, or to continue to do. Inability or unwillingness to provide clarity in this area causes people to become frustrated and confused and will either elongate the adoption curve or sabotage your program entirely. Since DG is new to your organization (assuming this is the first time you've tried it), you need a full-fledged education and training program to ensure that people understand and have the skill sets to be effective in their (sometimes new) roles. If you are making another attempt, you still need to review training and education to make sure you do not repeat prior mistakes. Make sure your program is comprehensive, moving from the general to the specific; the rationale and business case for DG to the specific skills or knowledge needed. While only a few people may need the specific information, the entire organization (and your implementation) will benefit from the "big picture" stuff.

Mandatory components of your change plan (regardless of approach—if you are planning change, you need all of these in the plan) will need to be:

- Specific conditions for sustainability and success
- Metrics defined for sustaining, effectiveness, efficiency, value (see next section)
- OCM team members, i.e. change agents
- Resistance-management plan-approved
- WIIFMs—Staff performance goals and reward structures to new accountabilities
- A sustainability checklist to monitor progress quickly
- Mechanisms for feedback from stakeholders
- DG communications plan-approved
  - Messages and branding—give your leadership something to say, because they have never talked about this before
  - Vehicles for communications—Be creative. Death by PowerPoint is a common side effect of DG
  - Timing, frequencies, and delivery means for ALL communications
- DG training plan
  - Audiences and type of training
  - Identify levels and extent of training: orient, educate, train
  - Identify vehicles for training
  - Define timing, frequencies, and delivery means
  - Review and approval of training plan
- Develop staff transition approach (use HR as needed)

> Remember, DG can generate a lot of resistance, and you cannot afford to ignore it. Resistance usually occurs across a continuum—from visible support and advocacy to overt hostility. Either end of the spectrum is easy to spot; but it is the passive resistors to worry about because they do not speak up and are very hard to spot. Make sure your plan contains the tasks and the time to understand and address all the types of resistance you will encounter. Your DG program will definitely benefit from the time spent. NEVER IGNORE RESISTANCE. It will not go away.
>
> Also, make sure your OCM plan gets integrated into the overall DG roadmap effort; these tasks do not stand alone, and it is essential that they be coordinated with the overall effort.

Ramification and benefits

Developing the set of tasks to make sure DG is sustained provides a level of proactive management of what it takes to get those critical behavior changes to stick. As stated earlier, many in the organization view things like individual departments with mission-critical spreadsheets as perfectly acceptable. Yet DG is viewed as a roadblock to getting things done. Therefore, the goal with a good change management plan is to make sure that those types of perspectives are addressed and overcome. Then rapid adoption will occur, along with the potential for earlier benefit realization and minimal churn while the organization adapts to its new state. Do not underestimate the damage and cost of poorly managed change. It might not show up on the balance sheet, but it is there nonetheless—and it can be a huge number.

Sample output

Templates for the following are in the appendices, due to size limitations:

- Communication plan template
- Training plan template

Metrics

Don't look back!
Why not?
Because I just did! Run faster!
Terry Pratchett, Nation

Metrics and measurement are critical components of your DG program. By this time, some measurement has already occurred in the form of the change capacity assessment and the stakeholder analysis. Prior work areas have already implied metrics, such as monitoring understanding of behavior changes.

Regardless of approach—noninvasive or large profile—you need to prove effectiveness and value. You will also, most likely, be called upon to prove your "new way" of doing things is efficient, i.e. delivers value commensurate with the time and resources being applied.

There needs to be concrete and specific data that reflects whether or not the program and the change management process is achieving its objectives, and messages are being heard and training is effective. Value is being added. The approach to DG is effective. Can you prove it? In general, you want the metrics to:

Measure effectiveness

- Measure achievement of initiative goals and objectives
- Determine effectiveness of communication
- Determine effectiveness of education/training
- Measure value

Measure efficiency

- Measure efficiency
- Measure speed of change adoption

Each of these two categories of metrics are covered briefly in the following sections.

Define metrics for effectiveness

These metrics make sure that DG is getting accomplished in a manner that provides benefit to the organization.

- Measure achievement of initiative goals and objectives—generally stated in the business case.
- Determine effectiveness of communication—test how well key messages are getting across to the organization.

- Determine effectiveness of education/training—assess if the education has provided the target audience with the skills needed to be successful in the new environment.
- Measure value—What value is DG helping accumulate by ensuring data is well managed? This metric group will cover risks, returns (including various means of monetization), and data debt.

Approach considerations

There are many views of effectiveness. A good start is to define any metrics that indicate the goals of DG are being accomplished. (Another reason the alignment exercise we have covered will become important.) You can obviously derive metrics if data quality is a goal, or data literacy, or standardization.

Also consider the old stand-by list of People, Process, Technology, and Data. Are there any goals attached to these? If so, there should be metrics.

You will be challenged. IT is leadership's duty to ask, "Does it add value?" But that means you need to define what "value" means to your organization. Does the data "asset" get measured? If it adds value or is monetized, do I assign an income category? If it is an expense reducer, is it against overhead or gross income? How can data affect debt? How can data affect reserves, equity, or allowances?

The best way to further understand this category of metrics is to look at some examples of value metrics; metrics that show improvement or contribute to improvement of a balance sheet or income statement. They are listed by category of financial context, or people, process, technology, etc.

Financial context

- Operating Income by Knowledge Worker
  - Operating Income for year divided by number of Knowledge Workers
  - Knowledge worker is defined as someone who uses information to make decisions and take actions that cause the fulfillment of objectives, reads information

Areas of data monetization

A good source of metrics to see if data management and governance is working is to categorize business initiatives by the usage of data. This table was originally in the alignment section of the companion enterprise information management (EIM) book to this. "Making EIM Work for Business."[1] The original intent was to provide a thought starter on how to monetize data. Obviously, that has caught on. DG capabilities are required now for advanced analytics and AI. But other uses of data can also "monetize data." Consider this list of categories of using data to improve business outcomes. Any DG (or DM) activity in support of these can be measured as to application of the DG or DM activity to the amount of benefit received. You will definitely get metrics that can be used to show the value of DG (Table 10.1).

Data debt

Another financial area is data debt. This is not a real number (yet). It is a way to communicate the cost of mismanaging data, stated in terms of an obligation to the future to fix the data problem. "Data debt" is a term taken from the Agile Development world and the concept of "technology debt." I ran into the term while with First San Francisco Partners when a client asked, "Is there such a thing?" Data debt is

---

[1]Ladley, John "Making EIM Work for Business" Morgan Kaufman, 2010.

Table 10.1 Data monetization areas

Usage value category	Data, information, and content used to improve or achieve goals	Consider these types of metrics
Processes	Improve cycle time, lower cost, improve quality	What DM and DG capabilities are needed to monetize by saving money, and does their execution coincide with saving money?
Competitive position	Capture competitive intelligence and differentiate yourself	What DM/DG capabilities are needed to use data to create a differentiator?
Product	Create, package, and market unique, higher margin products	What DM/DG capabilities are needed to use data to identify a new product or feature?
Asset/intellectual capital	Prolong leadership, embed knowledge into products and services	What DM/DG capabilities are needed to use data to embed knowledge into a service?
Enabler	Foster employee growth and empowerment	What DM/DG capabilities are needed to use data as a means to enable employees to do better work?
Risk manager	Manage risk, of various types, that threaten value by increasing liability	What DM/DG capabilities are needed to use data to reduce risks?

a concept and metric that will reveal to leadership the huge costs in delaying doing the "right things" with data and information. Organizations accumulate massive data debt by mismanaging data, e.g. they decide to create the short term, nonintegrated data table, rather than modify an existing data store. The longer you wait to fix data, that is, develop a well-governed architecture, you will rack up future costs to correct it. Other examples (metrics) of accumulating data debt are:

The increasing cost to clean up poor data quality.
Excessive costs from misaligned BI, i.e. doing three projects to deliver the same data to three departments.
The cost of not being able to count how many customers you have (or items, or products).

Mismanagement of data creates a debt accumulation situation. The longer you keep doing dumb things with data, the more expensive or the bigger disaster you will have when you need to or want to fix it.

Think of DG as the data debt repayment process.

Risk

Risk, as a financial metric, represents financial risk that could happen or has happened due to inadequate or incorrect data. Some examples of this type of metric are:

- Threat metrics
  - Cost per downtime event
  - Loss of customer confidence
  - Civil action via lawsuit

- Financial risk metrics
  - Liquidity
  - Operational costs
  - Equity/market value reduction
- DG compliance
  - "Hits" on web-based tools
  - Access counts on repositories
- Legal compliance
  - Potential penalties per subject area
  - General fines and penalties
  - Litigation fees over time

Metrics for meeting DG goals

We can measure our effectiveness by seeing that the people, process, technology, and data goals that a DG program establishes are being attained. *Some examples of this type of metric are:*

People

- # of data owners identified
- DG adoption rate by company personnel (Survey)
- Productivity increase due to efficient issue resolution

Process

- # of DG decisions backed up by the steering committee
- # of approved projects from the data governance working group (DGWG)
- # of issues escalated to data governance council (DGC) and resolved
- # of data consolidated processes
- # of approved and implemented standards, policies, and processes
- # of consistent data definitions
- Existence of and adherence to a business request escalation process to manage disputes regarding data
- Integration into the project lifecycle process to ensure DG oversight of key initiatives
- Increased efficiency of projects and new project initiation/reduced costs
- Communications events that happened as planned
- Training events that happened as planned

Technology

- # of data sources consolidated
- # of data targets using mastered data
- # of spreadsheets used
- Data integrity across systems
- Improved traceability of data
- Usage of a Unique Identifier
- # of business terms mapped to data models and objects
- % completion of attributes

- % completion of lineage
- % completion of glossary
- Improved reporting efficiency and accuracy

Ramification and benefits
Some metrics, as you can see, are very basic. You need to use the ones that help you show effectiveness in a way the resonates within your organization.

Define metrics for efficiency
Efficiency metrics considers the effort being applied to DG and concerns itself with making sure you end up at the net zero level or show a return on investment. Some of the effectiveness metrics can form the numerator or denominator of efficiency metrics.

Approach considerations
There are a few categories available for efficiency metrics. One is total cost of ownership of data. The other is the efficiency of the operating model, i.e., is it useful?

Total cost of ownership
What does data cost an organization? Surprisingly, a lot more than most CIOs or CDOs believe. If DG is to ensure effective use of data assets, then DG needs to understand costs as well as benefits of data.

People-related efficiency metrics

- % employees using BI and reporting tools
- % departments head counts that are business or data analysts
- Total staff cost for direct BI/DW work in IT
- Total staff cost for direct BI/DW non-IT (departmental or center of excellence)
- Total labor cost for Business Users of BI and Reporting
- Total labor cost for support of Business User BI—if different from non-IT
- Total labor cost associated with BI/DW
- Total labor cost/business BI user; if overhead numbers are available include those
- Total labor/All BI users—Divide total of all BI reporting labor by number of users
- Total BI/DW cost per Knowledge Worker (KW) (includes infrastructure and licenses)—IF BI users and knowledge workers are materially different, then develop a cost per KW same as you developed an income per KW. This can allow for some interesting rations.

Data related metrics—i.e. data movement and maintenance

- Number of files sent downstream—for example, the number of .CVS files that are provided to departmental users
- Cost per non-DW Interface—What does it cost to extract data from a system to provide a file to users. Most organization do this, and most have no idea how risky or expensive it really is
- No. DW files, cubes and marts accessed by users
- Cost per DW Interface—What does it cost to extract data and create a file from the DW?

- No. of MS Access databases—this can be a staggering and scary number
- Data Redundancy (Duplication of data elements across the entire data landscape—this can be time consuming and I only use it for the most nasty old "spaghetti" architectures.)

Data portfolio management—What are the costs and metrics for managing the data portfolio?

- Add new data to "EDW"—How long does it take to add a new column of data to the exiting DW? (not just create a new table for a report—that is cheating)
- Subject Areas supported (as integrated data)—what domains or subjects are represented
- BI or formal Data Strategy in place—simple metric—Yes or No
- Data at Risk (Personal and other regulated data)—Identify data that can create risk from misuse, poor quality, or compliance issues. This becomes a metric when you assign a possible cost, due to impact or fines.

Technology

- BI Software Vendors
- BI Software License Maintenance per year
- DW Maintenance budget
- DW infrastructure budget

Process—Operating model efficiency

- Policy responses
- Issue resolution

Ramification and benefits

Obviously, there is no shortage of metrics. The key is not quantity—it is starting with a few that tell the story you need to tell.

Define metric collection and reporting

Of course, regardless of what metrics you need or want to produce, you need the means to produce and report them. This section represents activity to be used when you need to figure out how to gather the data for metrics.

Approach considerations

Some metrics are gathered by observation. That is, how many people showed for a meeting? These are easy. Reporting them is a matter of a presentation of update to the DG web page.

Others, like income per KW, require understanding where to look, and how to count. You will need to defend these, so the methodology is important.

May metrics will have an automated source of metrics as DG tools get more sophisticated, and you can report on what data elements were used from the glossary, or how many times a tool was accessed.

Ramification and benefits

Reporting metrics is best done through a web site of some sort. I have seen many varieties. But if you are going through the trouble to collect data, then invest a little more time and make sure it can be easily accessed. A periodic review of some numbers is nice, but people forget the numbers in between meetings. Making a scorecard available where anyone can see the value of DG at any time is a huge boost to the program.

Implementation—Rocky Health

Tom has to address the common issue with many DG programs of having to make a second go, or evolve a prior program. This means there are some good feelings and some bad feelings.

The Implementation work area is where all of the "get started" activities live. The implementation activities, therefore, must not only produce the list of events required to deploy DG, it must also provide an outline for success and sustainability.

There are ample opportunities to implement changes across fundamental DG capabilities. Since incremental changes have worked well in this culture, Tom will recognize it is a good idea to continue this. In addition, the prior success was based on a new policy and procedure—the issue reporting. The capabilites point to simialr steps to take. Standardizing metrics, and new accoontability for data stand out. Since new accountability is a big cultural shift, Tom can elect to focus on standardiing metrics and reports.

Of course there will be other activities—projects to suppoort use cases, support existing projects, policy roll-out, new technology, etc. But these are all over time. Tom's focus will be on an Agile-type implementation, with bursts of activity that will incrementally move the organization forward.

The roadmap should specify various tracks: Projects, processes, people changes (although there are not many of those initally), and any possible technology.

Tom decided to start with another policy—since standarizing metrics is important, he creates a process similar to issue reporting, where new reports, dashboards, or request to measure anything need to be submitted on a form. This is a bit riskier than offering to solve people's isues, but that is the first DG process to implement a roadmap that follows with additional items, a summary of which appears below as Fig. 10.10.

Tom also realizes that, even though he will continue to do small tasks, some training and communications will be necessary as the visibility of DG increases. So some sustaining requirements are developed. Ideally, this should have been done sooner, but Tom already knew the culture accepted a low profile approach.

A small set of metrics are needed—mostly participation in the new processes and merasurement of the accuracy and efficiency of reporting.

Implementation—Rocky Regional Electric Coop

Diana has a much more complicated problem. Every effort she needs to govern is visible, and top priority to the respective stakeholders. Ideally, a culture assessment would have been done sooner. If not, it needs to be done now. In the United States, companies providing energy to households are highly regulated and under rigid scrutiny from state and federal regulators. Culturally, they can be extremely inflexible. Besides an intricate roadmap, Diana needs to identify a thorough set of requirements for change management if they have not yet been done.

	2019		2020				2021				2022
	Q3	Q4	Q1	Q2	Q3	Q4	Q1	Q2	Q3	Q4	
Projects / increments to apply DG											
Predicitve analysis for scheduling	▨										
Outcomes analysis and registry changes—multiple sprints			▨	▨	▨	▨					
CMS reporting consistency			▨	▨							
Primary care metrics						▨					
DG—Implements new report and metric request log	▨										
On-going DG capability development											
Identify additional training opportunities	■										
Identify mechanism to store data decisions, CDE definitions				■							
Establish communication vechicle (web site or similar)				■							
New engagement and operating models				■							
Policy administration				■		■					
Data and metric standardization						■					
Track DG effectiveness						■					
Deploy DG scorecards						■					
Report on added value and promote data debt thinking			■	■	■						
Determine traceability and lineage requirements and strategy						■	■				
Transfer responsibility to sustain change to affected areas								■			
DG activites for each project, repeat by project											
Identify new capabilities or DG and DM											
Develop DG tools or templates for use cases											
Identify new critical data elements, metrics, or domains											
Define policy and standards for data related to projects											
Train relevant DGC members											
Oversee use cases and / or data issues											
CDE criteria for PROJECTS											
Define policy and standards for data related to use cases											
Use the operating model to monitor use case and oversee data											
Identify additional training opportunities for new DGC members											
Ensure new elements are "glossarized"											
Install new policies											
Communicate new capabilites and stakeholders											
Train new members											
Define new accountabilities and custodial duties											
Refine operating models if required											
Define roles and responsibilites											
Set policy											
Implement metadata solutions											
Manage DG metrics											
Track value of DG											
DG metrics presentation											
Define accountabilities and custodian duties											
Sustaining requirements or every project											
Leader alignment											
Revise and implement communications											
Data Literacy check											
Refine training plan as needed											

FIG. 10.10

Rocky Health roadmap

This is a good example of how the various work areas in this book are not purely sequential recipe steps. A best practice is to start organization change activity as soon as feasible. In the case of Rocky Health, later was OK. In the case of RREC, the sooner the better.

For her roadmap, Diana will not only need to consider the three identified initiatives but also look at any other initiatives that may have some sort of data affinity with the main initiatives. The DG capabilities need to be lined up with all of them, BUT they need to be broken into small pieces. There will undoubtedly be a short-term form of DG to get to minimum sustainable operating model (MSOM), but also an interim stage or two before any long-term vision is fulfilled. Fig. 10.11 shows the type of thinking process Diana will need to go through.

Sample capabilities	Capability increments	Sample Initiatives								
		Enterprise reporting and B1			Engineering & asset mgmt			Customer svc hook up		
		P1	P2	P3	P1	P2	P3	P1	P2	P3
Data quality	DQ1				X			X		
	DQ2								X	
	DQ3						X			
Metric management	MM1	X								
	MM2		X							
	MM3			X						
MDM	MDM1							X		
	MDM2								X	
	MDM3									X
Data standards	DS1				X			X		
	DS2		X			X				
	DS3									

FIG. 10.11

Capability and project increment chart

The roadmap will be substantial, but it is important to remember that few plans survive first contact. A multiyear plan could be intimidating. Diana's focus needs to be near-term success with some flexible long-term paths.

The sustaining requirements could be considerable. If we assume Diana has good cultural aware-ness, she will know where the resistance will start, and who will be the key players. Her resistance management plan will need to be approved by leadership. She can do a lot of low-profile projects, but she will still need a lot of air cover to get collaboration around data. A noninvasive approach is not an answer here, either. She can implement various aspects in a noninvasive manner, but full acceptance of DG across the initiative will require some leadership aircover. The issues won't be around bothering anyone directly involved. The issues will arise when a plant manager decides to not participate at all.

Metrics need to be applied judiciously, there are too many possibilities. She will need to pick a few key metrics to start that prove value and progress. She will also need to monitor for resistance.

Summary

There isn't a way things should be. There's just what happens, and what we do.
Terry Pratchett, A Hat Full of Sky

A rational roadmap is your penultimate deliverable. Any roadmap, even a low profile one, requires some sort of defined process so it can be defended.

There needs to be a conscious effort to maintain "sustaining" DG. Otherwise, DG will fall into the same traps as all previous DM efforts. That is why the discipline of organization change management is key. But the DG stakeholders also need to start right away. It is most likely a mortal blow to DG to train various groups to do something and then have nothing to do.

This activity is the first real interaction with the actual areas subject to DG. One concept that cannot be integrated into a tasks list or sample work products is the need to observe the actions and interaction of the DG team, stakeholders, and governed parties. To be very clear, somewhere along the way there will be eye-rolling, misunderstanding, or missed expectations. Metrics will go a long way to addressing the various types of pushbacks. There needs to be a keen awareness of what is going on behind the scenes, and a sensitivity to what is really happening versus what is desired to happen.

Essential questions

1. Sustaining requirements can be done sooner even though they appear in this section of activities. Why is that?
2. Why is a clear roadmap so important?
3. What is the long-term consequence if you do everything in a noninvasive fashion and never try and make DG a visible capability? Why would that be alright for one organization, and not for another?

Operation and change

11

Chapter Outline

Too many people were working on the mind without paying sufficient attention to the heart.
Kotter and Cohen

Overview

This section is about operating and sustaining the program. Operation of the data governance (DG) capability my not seem very exotic. However, experience has shown many organizations are often surprised at two things:

Data Governance. https://doi.org/10.1016/B978-0-12-815831-9.00011-4

1. There seems to be more work than anticipated. This is a function of getting the operating model in place, and learning curve. Remember that a proper DG program means some change has to occur, so there will be a startup period. But over time there should be less work as better data requires less data wrangling or duplicate efforts. Although you never want to advertise operating DG as a lot more work, remember we are looking to have the same or less overhead, so you do need to be honest about the simple fact that doing new things can put people off.

2. Even with adequate training, DG can seem "hard." This is a function of resistance and change. There will be new things to master, resistance to deal with, and assurances will need to be delivered to nervous managers.

Even low-profile efforts may experience some sort of operational start-up angst. Any time you make anything more formal, or install a new capability, there is a learning curve or slow response due to people learning that something is now "official."[1] In addition, there is usually some setup work for any tools, policies, or standards. Populating glossaries, installing technology, or getting a policy made "official" adds to the sense of additional work, even though it is one-time, start-up activity.

A common mistake happens during the big kick-off meeting (or little one, depending on the approach) setting expectations too high for a certain time frame. Another error is to introduce every single possible data steward or stakeholder even though they may not yet be sure what they are supposed to do.

Realistically, a DG operating model can take as long as 6 months to stabilize. Learning curve, resistance, and adjustments will all occur. When DG is implemented low profile or through projects, there still needs to be time and organization change management (OCM) activity allotted to educate and prepare the new participants. (Good project status reporting is essential here, by the way.)

This chapter will address resistance and sustainability in detail. It will also address a lot of activities that are generic and common to any organization, but review from the organization change management context. The format of this chapter needs to be different. The prior few chapters were oriented toward picking activities that are necessary for your situation, then understanding how to go about doing them. This chapter is more descriptive than prescriptive. Therefore, the sections are arranged differently.

Fig. 11.1 shows the types of activity in this area.

This area is a set of activities used to make sure that DG reaches "business as usual" (BAU) status. You are embedding DG capabilities (i.e., new business capabilities) into the organization.

Your mindset of embedding a new business capability is more than important. It turns out, it is critical. At this point far too many organizations go "Ok, we just need to do governance stuff." The attentiveness of the DG implementation team may fade a bit. This is not a good idea.

A new business capability requires much more than just "doing stuff." Remember, your organization is not used to this, or tried and it did not go well. You need to make sure the new capability is working the way it is intended. You need to HELP people move into a new way of looking at data and even their own work flow.

[1] Noninvasive and low-profile DG teams are often surprised. For example, if a very noninvasive effort requires a small change, such as using the formal process for getting access to a file vs an old informal process, there will be a sudden hesitation to select the new menu option unless you have anticipated that there will be concern or resistance. Think of what happened when iPhone iOS went from the old style graphics to new style. I didn't like it and I could find no rational reason, except it was different and no one told me.

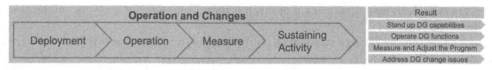

FIG. 11.1

Operation and Changes activity area

Fig. 11.2 shows that quite a bit of activity is possible at this point. If your approach is low profile, do not assume you will pick a few of these activities for your particular plan. You may have to be as detailed as a very visible effort if the desired operational state requires significantly different workflow. As you proceed through the material in this chapter, place yourself in the shoes of those who will be on the front lines of using new policy, processes, or tools.

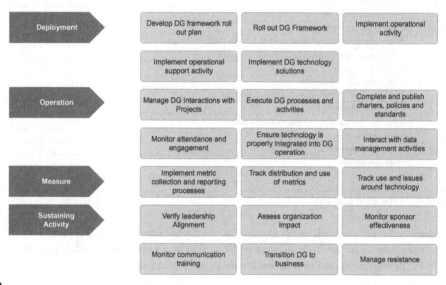

FIG. 11.2

Types of activity in operation and changes

Deployment

After struggling to get engagement, then designing and building a program, now you need to operate it. Deployment means choreography of several tracks. Low profile efforts may have fewer tracks, but it never hurts to do detailed planning.

Develop DG roll-out plan

Regardless of your approach, there will be multiple tracks—technology, data management (DM), OCM, operating models and framework, and implementation of policies and processes to operate DG.

Even if the individuals have done many projects, it is most likely some of the work will be new. The more detail, the better.

You can save time by having a standard DG roll-out activity checklist that is applied for every type of iteration. See Fig. 11.3 for an example.

Roadmap category	DG Expansion checklist, By iteration - Use case or project support	ITERATION	
		Q1	Q2
DG Ops	Add new use or project		
DG Ops	Identify new capabilities for DG and DM		
DG Ops	Develop DG Tools or templates for use cases		
DG Ops	Identify new critical data elements, metrics, or domains		
DG Ops	Define policy and standards for data related to use cases and projects		
DG Org	Train relevant DGC members		
DG Ops	Oversee use cases and / or data issues		
DG Org	Modify DG charter if required		
DG Org	Identify additional training opportunities for new DGC members		
OCM	Communicate new capabilities and stakeholders		
OCM	Train new members		
OCM	Realign leadership if required		
OCM	Hold Town Hall or other cross-functional meeting		
OCM	Realign DGC if required		
OCM	Define new accountabilities and custodial duties		
OCM	Refine operating models if required		
DG Ops	Ensure new elements are "glossarized"		
DG Ops	Install new policies		

FIG. 11.3

Data governance expansion checklist

Depending on scope, your roll-out plan may cover only the narrowest of DG capability roll-out. Or the roll-out plan may cover multiple aspects of DG, such as an operating model, new services made available to the organization, and new technologies to implement and use.

Roll-out DG framework

You can engineer the managing activity for DG but, until it is deployed, you can never be sure it is ideal. An operating model is not like a light switch. Since they are almost always virtual communities, declaring operational status means a lot of communicating. Defining the management aspects of ongoing DG and specifying some immediate activity goes a long way toward a more refined and sustainable DG process.

You can also adjust other aspects of the DG operating framework as well as the charters as you start to see the operational details unfold and you define the initial DG tasks.

Operating frameworks can be implemented top down or bottom up. A low profile effort is usually bottom up, calling in leadership only when it is time to grow. A high profile effort, such as with RREC, will mean getting some "top cover" as well as installing lower layers to activate new capabilities.

Implement operational activity

Most important at this point is to specify what is being done on day one. Usually it will be training. But you need to plan for BAU activity. Remember DG is not process, it tends to observe. But occasionally there is stuff like glossary maintenance, etc. that needs to be started since it is the main focus of the effort.

Operational activity should be addressed separately from standing up the operating framework and operational support. For example, if you are on a second pass at DG you may have the operating model, and just need to start operational activity.

Best to have a checklist—refer again to Fig. 11.3.

Implement operational support activity

DG as a service (DGaaS) is a means to have organization adapt to formal data oversight. I don't mean you can outsource DG, but you can treat it as an internal service. So part of being operational is to start to offer support to other areas.

DG can assist projects and initiatives (initially) by maintaining the dictionary, assisting with data quality, and in the process, educating about DM and governance. DGaaS is a good tactic, but you will have some additional overhead to cover until reaching BAU status.

It is a bit contrary to a pure "V" but as I have said, the *pure* "V" is theoretical—it is an abstraction to help you build something with separation of duties.

Implement DG technology solutions

No rocket science here—few shops have never acquired technology and implemented anything, so a lot of the newness is normal. However, starting new things can be time consuming, and very few technology products install and stand up operationally without some sort of challenge. As of the writing of this book a lot of the DG tools offer fine features and functions but often have trouble inserting themselves into wide varieties of technical infrastructures—even cloud-based solutions have issues with interconnectivity. Make sure you plan for adequate time to install and achieve operational status before promising benefits that will be delayed.

Operation

I separated out operation from standing up operation so I could point out a few finer points. Once in operation, DG becomes a regular business capability area. But it is new, and I have noticed organizations overlooking a few items that seem obvious, but get lost in the abstractions and newness.

Manage DG interactions with projects

You MUST interact with project teams to get DG up and running. If you are lucky, a PMO will provide aircover. (Or should—if not find out about that while planning OCM or even the operating models.)

Even a noninvasive effort, which is treated often as a project, will interact with other projects. DG by definition interacts with other activity. Strangely, data people, who often lead the DG efforts, need some skill reinforcement on interacting with and tracking interactions across stakeholders and projects.

Execute DG processes and activities

Obviously, once DG is in operation, you need to do things like work with project teams, participate in planning, or whatever other process has been defined out of your capability requirements. If offering some DG services make sure that everyone feels confident in their assignments.

Do not forget that a lot of DG is interaction with other areas. Like compliance, privacy, applications development. Allow for suitable orientation and introduction for any new interactions.

Complete and publish charters, policies, and standards

Writing the charters and policies and standards is often the easy part. Occasionally stakeholders will seem surprised that a new policy is actually being seriously rolled out. A policy can sometime languish in an approval or prerelease status as a form of resistance, or if someone gets worried about negative feedback. Make sure the policy writers and other participants in this activity have a good process to get help in case there are obstacles or delays.

Monitor attendance and engagement

It's one thing to look around a room and see who made it to the meeting. It is another to track who and how often, record it, and report on it. It is really important you do the latter. Declining interest is manifested in attendance numbers. Resistance makes itself known by lack of participation as well. It sounds old fashioned but always pass around the sign-in sheet.

Ensure technology is properly integrated into DG operation

DG workflow will eventually interface with technology. Hopefully your operating framework or engagement models reflect the interfaces with technology. Even if there has been good training and your tools are working, always check on who is using the tool and when. For example, a steward may be accountable for some sort of maintenance in a glossary but has delegated the task to someone who can run a tool but is not fully versed in the DG effort. Also make sure the tool is being used according to the designated workflow and the tool is working at the level of performance you require.

Interact with data management activities

Prior sections covered that DM work is often from a DG directive. Implementation of a standard, execution of data profiling, use of a glossary, policy workflow, etc. This is the right side of the "V."

But that does not make DM a black box, or a hand-off. DG still needs to ensure that any DM processes are working as intended. Standards need to be monitored to make sure they are adhered to. I am very often surprised that DG assumes all is well, and DM activities are just happening, when often they get stalled or pushed aside. Never forget, at its essence, DG is the control of data and assurance data is managed.

Measure

Measures and metrics for DG (and DM) are often defined, but experience shows they are not often fully implemented. So if you are serious about measuring progress and effectiveness, you need to formally plan what you are going to do and manage the execution of measuring DG.

Implement metric collection and reporting processes

The earlier work areas had us define what to measure and design how to do it. There may be some development work to build reports, gather data from tool interfaces or web sites, or implementation of processes that simply count things. There may be a process to distribute results and ensure receipt and allow for feedback. There may be scripts or other coded functions to install. Either way, operational DG needs to manage metric roll-out as a subset of the entire program.

Track distribution and use of metrics

Hopefully you have designated who will get the metrics. If you are emailing reports, or simply providing an alert that a metric is available for viewing, this task covers the monitoring of those processes. You also need to ensure that metrics are used, that is, the report isn't set aside, and proper action is taken if the measuring of DG indicates a problem.

Track use and issues around technology

It's very common for new technology to exhibit growing pains. Even if the tool is best-in-class and widely in use, there are always issues learning the idiosyncrasies of tools. There are always adjustments to workflows. Therefore, monitor deployment and use closely. Do not give anyone an excuse to resist over a tool. Deal with tool issues rapidly. Many data efforts of all sorts have been grounded by so-called tool issues. Sometimes it is a legitimate problem with the tool. Others, it is a lack of patience to work things out.

Lastly, be very candid with tool vendors and insist on full support and responsive behavior. At the time of the writing of this edition, there are many tools being deployed to support DG, and some of the tools are still maturing. I have seen vendors service the large, big-name clients, and give short shrift to smaller organizations. Also, rapid growth in a technical sector means service and support skills trail delivering features. Be very firm with your vendor to get the support needed. Do not accept delays. In this hyper-connected social media world, a small company can make a very large deal about a vendor. Use whatever leverage you can.

Sustaining activity

One of the more important aspects of operating any sort of DG is making sure it "sticks." There are countless examples of program that seem to fade into oblivion. They may have had the right approach, the right team, and had the right capabilities in mind to add great value to the organization. But without a deliberate plan to sustain the program and subsequent execution of that plan, the programs fade. Of course at the heart of sustaining the program is OCM, which you will have.

Most of the work I have done with second and third attempts at DG were at shops that did nothing to proactively sustain their programs, assuming that they would sustain themselves. This cannot happen. Even the smallest, most noninvasive efforts still need to change something, and there may be something in that effort that causes resistance. Or, there just need to be good communications and training.

Regardless of your approach, you must manage the people. If you do that, usually the rest of the program will work itself out.

Verify leader alignment

The mere thought of this activity frightens data folks. It seems to be an overreach. Who are we to tell leadership what to do? However, it is not that kind of activity. The purpose of this activity is to ensure that your leadership shares a common view of the DG program. Actually, this technique can be applied to DM or any other program.

This needs to be done so that everyone is, obviously, of the same mind as to the drivers and approach to DG. But also, to confirm that everyone understands their role. This is important to do as soon as possible in the operational phases. Ideally, this activity is done a lot earlier. A version of this should be done in Engagement do get leadership to participate during design and strategy steps, but at this point you are absolutely confirming their state-of-mind. You are looking to make sure they understand business value, how to walk-the-talk, and promote the program. You will also need to repeat this activity once a year, or before incremental releases of program capabilities.

If you are reading this and are on your second or third go-around of DG, you may be reflecting on this, as in "were my leaders on the same page?" The answer is often "no." When they are not aligned, then there is built-in friction, and when resistance crops up, the friction just intensifies any uncertainties.

The outcome of this activity depends on how aligned leadership is. When you do this earlier in the program, they will not be aligned, but participation and education will line them up. At this point any misalignment will need to be quickly corrected by the sponsor.

Here is a typical list of questions that can be asked to determine alignment:

- What do you think the ultimate contribution of DG will be for our organization?
- From your perspective, what will be different when DG is fully operational?
- What is your definition of success for the DG?
- What are your biggest concerns about the changes that DG will drive? How would you address those concerns?
- What do you think your role is as a leader in making DG a success?
- Considering upcoming resource and effort requirements to carry DG forward, how would that fit into how the executive team sets and manages priorities for the business?
- Who is accountable for delivering on DG results?
- What do you think are the best ways to encourage positive reception of the change by key stakeholder groups inside and outside of the organization?
- When considering the upcoming DG effort, what do you consider an appropriate strategic planning horizon (length of time) to consider?

Monitor sponsor effectiveness

Similar to leadership, the sponsor needs to be monitored when operational. It has already been stated that the sponsor is very important, regardless of approach. They are the main impetus of change and need to reflect the importance and priority of managing data assets. Since very few sponsors understand that there is an art to be a sponsor, very often the DG sponsor is not as effective as the program requires. Assessing sponsor effectiveness is very important. In addition, as a program matures, you may need a new sponsor. Resources come and go, so it is not rare for an DG program to have more than one sponsor over time, and each one needs to be monitored for effectiveness.

A lot of research shows that the sponsor is a top success factor in making any kind of change happen. Prosci, Bridges, McKinsey, etc. have all done good work in this area.

The sponsor is the source of authority. For DG that is critical. They will:

1. Make sure issues get resolved in a timely manner
2. Evangelize the program
3. Build support for change
4. Make sure resistance is addressed
5. Manage expectations
6. Align and build support among other leaders

Your sponsor needs to participate, not just during Engagement and other preoperational activities, but after you start governing stuff. They need to be visible.

Table 11.1 below shows a typical sponsor plan:

Table 11.1 Sponsor plan	
Sponsor activities	**Audience**
• Engage other leaders in support of the changes • Reinforce the role of the project sponsor throughout project life cycle • Demonstrate active and visible participation throughout the entire project • Review the roadmap often. In light of DG progress, what needs to be added or changed?	• Peers
• Develop the activities by which the sponsor will communicate with the project team • Frequent interaction with team—status, issue resolution, vision	• Project team
• Identify actions that the sponsor will take with leaders/managers to ensure support for the project objectives • Use current organization charts and color-code: green for support, yellow for neutral, red for resistant • Find the opportunity for one-on-one meeting with those you have identified as resistant. Ask questions to learn the reasons for resistance. Be clear with expectations for the future	• Leadership
• Share the project objectives and key messages with employees • Reinforce these key messages at every opportunity; tell them what they should expect to see happen and when; link the changes to the data governance future vision • Be vocal and visible; provide opportunities for employees to communicate directly with you • Don't sugar-coat what will happen. Make it clear you understand the difficulties of going through change and that you identify with the learning curve and newness that may be experienced	• Staff

Sponsors also need to be assessed, and like leaders, as soon as possible. But during operations, it is essential to periodically take an objective look at sponsorship effectiveness. A really good sponsor will self-assess. See Table 11.2 below for an instrument that has been handed over to sponsors for them to make sure they were up to the task.

Table 11.2 Sponsor evaluation	
Sponsor characteristics	**Sponsor thought process**
1. Has a clear vision for data governance what it is trying to accomplish	Has a clear understanding been established of what we are trying to accomplish with data governance and why we are doing it?
2. Ensures that leaders and managers in surety endorse data governance	Who else besides me needs to endorse and support data governance?
3. Is aware of potential resistance points in the organization	What and where are the "land mines" that could derail data governance?
4. Understands the implications of data governance for people in the organization	How are people likely to respond; what must be done to build support?
5. Demonstrates perseverance when challenges arise	What obstacles and challenges are likely to emerge and what should I do when they occur?
6. Maintains involvement and takes ownership for the program	Will I be available to provide input and make decisions when required?
7. Provides visible support	What actions can I take to demonstrate support for data governance?
8. Holds people accountable	How will I hold people accountable and encourage the right behaviors?
9. Communicates effectively	What, when, and how will I need to communicate to our leaders, staff, external partners, and others about data governance?
10. Maintains momentum	How will I be able to assess and ensure continued progress?

Here is a list of characteristics that will help assess your own sponsor's effectiveness.

- Knowledge of change management processes and principles
- Understanding and support of the DG program
- Able and willing to be an active and visible sponsor of the change
- Experience and success rate as a sponsor/business lead of past change projects
- Ability to communicate the vision and need for change to employees and managers
- Degree to which the organization (employees and managers) would listen to and respect communications and support from this business leader
- Ability to influence and build support with other business leaders
- Ability to provide resources and funding for the project
- Degree of direct control this sponsor/business lead has over the people and processes being impacted by the change

- Degree of direct control this sponsor/business lead has over the systems, and tools being impacted by the change
- The level of awareness sponsors have of the importance they play in making changes successful

Space does not permit presenting more of the wealth of material that is available on sponsorship, and change management in general. Since sponsorship is so critical, I recommend Prosci, Bridges, and Kotter in the change management world as sources for more sponsor material.

Monitor communication and training

The most abused deliverables are training and communications plans. Once the program starts, for some reason many shops just stop looking at these plans. Inevitably they find out that communication events are slipping away, or training events are not occurring when scheduled. Even the smallest increment of DG will need some sort of communications or training. Print off whatever it is you develop. Enlarge it. Make a poster. Paste it up in some cubicles, or in a hallway. Check things off. If you are an Agile shop set up communications and training sprints.

I know this sounds really trivial, but I cannot emphasize that it is silly to go through the effort to define all kinds of events and not do them. If, for some reason, your program was suppressed and you were told to delay certain events.

Transition DG to business

Since DG is a business capability, as some point, you will need to get DG to officially move away from data and technology people and establish a DG presence as an official business capability. By "business" I mean organization in general. So this also applies to nonprofit, NGO, government, etc. Transitioning from program development and operation to on-going business capability is a common step in all sorts of major change efforts. This usually happens as part of compliance or risk management areas but could also be part of a line of business where data is heavily monetized. Either way, this represents recognition that DG is BAU. Since most readers of this edition are working in programs that are low profile, or just stating, this transition may seem far into the future. Rest assured, I have seen many organizations achieve this status.

Manage resistance

Resistance is defined as any activity that acts to slow down, question, or even prevent, your program from getting started or continuing. Some form of resistance is inevitable unless there is a significant burning platform, or your initial DG effort is so low profile that no one notices. Participants in change efforts exhibit various types of behaviors. In simple terms, there are three broad types: supporters, neutrals, and resisters. Typically, it is a bell curve, as shown. And like any other population, the 20% can make things miserable for the 80%. See Fig. 11.4.

Obviously, anyone who is less than supportive will require some sort of planning to identify the resistance and to deal with it. However, you also need to address supporters as well.

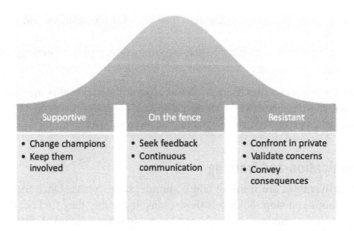

FIG. 11.4

Types of resistance curve

An effective resistance model is one that has the following five types of participants.

1. Active support/advocacy—These participants are often your evangelists. If they are in leadership, they easily permit resource reallocation. As stakeholders they embrace new things and carry the message to their peers.
2. Passive support—Someone in this group certainly desires success. But they are shy. They show up to offer support, or will evangelize when asked. This type of stakeholder embraces new things but only moves out of their comfort zone if others are doing so, OR if specifically asked by a supervisor.
3. No resistance/no support—As shown in the bell curve in Fig. 11.4, most of your stakeholders will be in this group at one point or another. If they are participants they may have been appointed. They rarely contribute, but also do not get in the way. Since this is a large percentage of your audience, they cannot be ignored.
4. Passive resistance—These participants are the perennial candidate for being the most aggravating. They nod in agreement, and when they leave the meeting they are planning how to not do what they committed to, and make it look like someone else is to blame. Or they are playing politics, and trying to please everyone they interact with. One symptom is a stakeholder who says they will get on board with the changes, but not until they get something else done. They are using prioritization to delay dealing with the change. Resistance also takes the form of repeated questions, and repeating the same question, or seeming to not "get it." These stakeholders will often manifest uncertainty by foot dragging, or asking for more clarification than is reasonable.
5. Active resistance—These are easy to spot. These stakeholders do not even participate, and initiate all conversations with "that will slow us down." They roll their eyes in the kick-off meetings during the engagement activity. They visit the sponsor or go over the sponsor's head with their opinions. They will even vocally and deliberately disregard new processes or standards when they feel they are unnecessary. Often, they feel they are too important to be chastised for not participating, because their other work is too valuable. Sadly, I have seen situations where an active resister has been allowed to continue because of a perception that their other work cannot be disturbed. Of course, this can derail an entire program.

There is an essential set of tactics that must be used to manage stakeholders, regardless of type of resistance:

- Learn what is important to each person and build support person by person
- Move people along one notch at a time
- Don't create unrealistic expectations—no change will resolve every problem
- ALWAYS FOLLOW UP ON YOUR PROMISES TO GET ANSWERS—IF YOU DON'T, YOU WILL SHATTER TRUST AND CREDIBILITY
- Do not make commitments that you cannot deliver on
- Pay attention to team members who have the strongest influence over other team members
- Establish clear vision for the change and a plan to drive it
- Communicate early (and often) to affected stakeholders to educate and engage them in the process
- Provide opportunities for and responsiveness to feedback
- Offer effective education when needed
- Align policies/practices, rewards and recognition, and organization changes

Beyond these essential tactics, each type of stakeholder has their own set of tactics you can deploy. Table 11.3 shows what to do for each of these resistance types. Again, space prevents me from showing a lot of examples, but look on the internet. There is plenty of help.

Whatever you do to address resistance, there is a short list of DON'TS:

Table 11.3 Tactics to address resistance	
Active support/ advocacy	• They often appreciate being kept informed and asked their opinions. When possible, share information with them and solicit their input to keep them motivated and involved • Ask a senior leader to acknowledge their accomplishments and communicate that they are valued • Solicit help in identifying the readiness of others • Ask for ideas to help move the rest of the team along • Encourage them to voice support publicly, explaining how it will help others • Ask them to help with someone who is struggling with the change • Regularly thank them for their enthusiasm (once is not enough!) • Don't assume that they will remain active supporters without attention; they can slip down the continuum if ignored. People become most susceptible when those initial inevitable problems arise, and their high expectations are not fully met
Passive support	• Acknowledge them when they demonstrate the right behaviors in support of the change. Sometimes positive reinforcement is enough to move them to active support • If you feel someone "gets" it but isn't voicing support, ask him to help another who is "on the fence." Sometimes voicing support helps an individual move to a more active support role • Identify opportunities for them to be more active in supporting the changes, perhaps asking them to describe something they did that really worked well. Position them as the "go-to" people for a particular topic or issue • Recognize that some people aren't comfortable voicing support publicly. They can still contribute positive momentum by doing the right things • Spend time directly asking for assistance and indicate how important they are. Also bring in supervisors if necessary

Continued

Table 11.3 Tactics to address resistance—cont'd

No resistance/no support	• Find out what is keeping them from being more comfortable with the change. Sometimes asking a direct question will work: "What would make you more comfortable with the change?" • Ask them what they like and dislike about the change, and what would make the change more successful • Tell them about changes you've made because of their input • Encourage them to come to you and ask questions about the change • Reinforce why the change is taking place and the benefits of the change for key constituents • If they don't believe that the change will impact them, educate them on how and why their roles may change; clarify expectations
Passive resistance	• Assuming there is adequate education being offered, these individuals require clear goals and WIIFM to be willing to get through the period of uncertainty • This situation requires leadership to reaffirm or shift priorities • Education needs to be adjusted to ensure that the necessary concepts have been conveyed and understood. This means measuring understanding, e.g. using quizzes and tests after training classes • Try and get peers to influence • Identify who they are. How? Since they won't verbalize their concerns to you directly, one way is to ask active supporters to help identify those who are struggling or dragging their feet • Schedule one-on-one time with them in private, since they won't express concerns in a public setting. Do not ignore them • Encourage them to voice concerns by explaining that it's natural to have concerns. Engage them in discussion of what would work better. A key first step is to get them to acknowledge their concerns • Connect actions taken to concerns they voiced and make sure they know why you did it. This can be a very effective way to move them along the continuum • Make it safe for them to voice concerns. Many times, their reluctance to do so comes from being beaten up for voicing concerns in the past • Be clear about nonnegotiable items but identify where input is possible and actively seek their input • Acknowledge and empathize that change is disruptive but continue to emphasize the reasons and benefits of helping the change be successful • Be consistent and steady in your own support for the change • Ask for direct help from leadership if necessary
Active resistance	• Ensure feedback mechanisms are working and understood • Give them the opportunity to vent without shutting them down. That will only drive them underground • Demonstrate understanding of the concerns even if you disagree. You can empathize without implying that all their concerns are valid • Don't shift blame elsewhere. While it may make you feel better at the time, it only confirms their concerns • Because they are visible and vocal dissenters, they tend to monopolize your time. Don't spend all your time with them to the detriment of others • Identify at least one concern that is valid and actionable and try to address it • If you take action, tell them what you did and why • Be clear about what is nonnegotiable and identify where input is wanted. While it's ok to let them vent a little on those nonnegotiables, encourage them to focus on what is within their control to influence • If you form a working group, invite an active resister to be a member. You will gain an important point of view and engage one of them in a positive way • Some may be responding because of prior negative experiences. Try to learn about those and clarify what's different this time • Mitigating this situation requires leadership to reinforce the direction and emphasize any mandates that have been made • Face-to-face conversation to identify root cause of resistance

Don'ts:

- Ignoring resistance and expecting it to go away—it usually gets worse
- No planning for resistance to occur—it always will
- Not listening to and understanding the concerns and feedback of those impacted
- Concerns do not always equal resistance
- Even those who resist may have valuable insight
- Inconsistent and ineffective communication

Operation and changes—Rocky health

Many things are possible once Rocky Health begins to expand and operate new iterations of its DG program. Here are two common scenarios that often happen to programs as they expand.

1. As Tom expands DG policy, a few departments do not participate fully. In one instance, Tom indirectly finds out that a few business analysts in the surgery department have created a set of reports and a small dashboard to monitor infections after surgeries. While this is well intentioned, wellness is an enterprise goal, there are other reports and databases to fulfill this report that are already in production.
   a. Tom needs to find out why this is happening. His response depends on the reason. If this is resistance, then he needs to address the source. However, there could be an issue with education and communications.
   b. If resistance, he needs to get his sponsor and the department head of the offending area in the same room at the same time. This is why sponsors are so important. If the sponsor fails to get the other department head on board, the entire program becomes not enforceable.
   c. If communications or training, Tom needs to find out where the gaps are, and immediately apply some remediation.
2. A few months after Tom launches the new DG program, Rocky receives notice of a $US14 million fine from a regulatory body. This is a significant fine and was based on data reported in error. The Chief Medical Officer promises regulators it will not happen again and acquires a reporting package from a vendor that provides the specific report to the regulator. Tom needs to address this issue quickly.
   a. First he needs to gather the data to show the impact of the new package—any data risks, new overhead, data consistency or timing issues, etc. He then needs his sponsor's help, to present the impact of the decision. Tom's leadership team will need to determine how to proceed. Either way, this is an excellent opportunity for Tom to present the concept of data debt. The organization is already accumulating costs to deal with the scenario.
   b. While resistance is a possible explanation, normally these types of event are a result of a type-A personality believing this is a special situation. This is usually a data literacy issue. Many executives need to see DG address these types of problems before they "get it." Additional education is usually very useful.

Operation and changes—Rocky regional electric coop

Diana will have a fairly robust set of requirements to sustain the REC DG program. Her training and communications plans will need to reflect a growing number of stakeholders over time.

A few features of Diana's operational program that would not appear in Rocky Health would be:

1. More varied training plan to reflect multiple audiences.
2. Communication plan reflecting many types of communications vehicles—posters, town hall meetings, as well as traditional training and education events.
3. A stakeholder analysis with predicted resistance areas and individuals. Her industry is well-known for episodes of inflexibility.
4. Integrated DG support with the Enterprise reporting and business intelligence effort, focused on the two capabilities of metrics and data standards.
5. Integrated support with EPRI effort, also using the shared work on data standards.

This all prepares Diana for issues and resistance. It will not prevent them.

For example, within a few weeks of kicking off the support of Enterprise Reporting, the internal team assigned by AppDev begins to complain. Historically, AppDev has been very responsive to delivering anything any user asked for. Of course, this included multiple version of metrics, reports, duplicate technologies, and all of the other items that compound data debt. They are not used to getting their own customers to align to standards. Diana can do two things based on her planning:

1. Ensure all stakeholders have received orientation and education. Once that is confirmed, she can address their hesitancy as passive resistance.
2. She needs to get closer to AppDev and provide support on all definition of metric activities. This is DGaaS, and while not perfectly a control function, it will smooth the transition for AppDev.

The next issue she confronts is pure resistance. A plant manager attends an introductory orientation for how the engineers are going to transition to a common set of data meaning in support of configuration and equipment management, and flatly says "my people do not have time for this. We are not participating." And leaves the room. Hopefully, Diana is prepared for this by understanding that this individual was likely to do this. In that case she will:

1. Get her sponsor to break open the resistance tool kit and seek high level support.
2. Start to build the cost to the organization if this individual maintains a nonstandard plant.

Planning ahead allows leadership to visit with the plant manager and offer some remedial education and reestablish priorities. The business case for the new standard will be useful as well.

If Diana did not prepare, she will need to present with the business case, data debt potential, and intangible impact to DG leadership and ask for an intervention.

Summary

The essence of Operate and Sustain activity is exercising the capabilities of DG, the execution of OCM, and making sure both are working properly. "Changes of any sort—even though they may be justified in economic or technological terms—finally succeed or fail on the basis of whether the people affected <u>do things differently</u>."[2]

---

[2]William Bridges, *Managing Transitions* (Cambridge, MA: Perseus Books Group), 2003.

The change management items address the people aspect. Getting people to address the emotional element of adopting discipline that has been nonexistent. "More than any other single finding, we discovered in this … project that people changed less because of facts or data that shifted their thinking than because compelling experiences changed their feelings. This emotional component was always present in the most successful change stories and was usually missing in the least successful. Too many people were working on the mind without paying sufficient attention to the heart."[3]

DG needs to be applied to various programs and projects. This means integrating DG with these efforts. A PMO is very helpful as the vehicle for DG. If none exists, then the sponsor and operating framework needs to make sure that individual efforts are using DG oversight.

Follow your training and communications plans. They are there for a reason.

Lastly, the various programs and projects being governed need to be monitored for the effectiveness of the DG processes. Frequent collection of metrics and successes of the data-related projects is essential.

Essential questions

1. Is it ok to try and predict how people will react to changes from DG?
2. If you vary from a pure division of governance and management, you affect checks and balances and DG does not work. Please explain if this is true.
3. Data debt can be a good motivator for data governance, but as a periodic measurement it is too abstract. True or False and explain.

---

[3]John P. Kotter and Dan S. Cohen, *The Heart of Change: Real-Life Stories of How People Change Their Organizations* (Boston: Harvard Business School Publishing, 2002), Kindle edition.

Final items and summary

12

Chapter Outline

The longer you delay treating your data as a significant asset, the harder all of your business decision making will become. If you ignore treating data as an asset, your organization will become unmanageable and never attain its potential.

The Author

The final words of this book will, and should, be a summary of what has been presented. But before we recap, it's important to reinforce that implementing more discipline in regard to data, information, or content is not really an option. As we have seen, there are a variety of paths toward data governance and management, and this book, hopefully, made it easier to determine your particular journey.

An all too common scenario in my career is an information management–related project for a new client—with one of the requested deliverables being "data governance." A request for it is given to the consulting team by the CIO, who was soliciting the consulting firms. To be precise, the request is usually to "deliver a comprehensive data strategy, with recommendations on ETL, MDM, BI, and data governance. Please include an ROI." The blast of acronyms is the first warning sign. The lack of any mention of the business is the second.

My teams usually have someone familiar with the industry or business model. The typical proposed approach is based on an assessment of the current maturity and business alignment followed by a data governance (DG) strategy—using the tools and processes presented in this book. Then we discover how hard this new gig will, or will not, be. If the CIO says "I know the answer, you just need to give me a list of vendors," then we are in the high risk zone. Often, when we attempt some education that DG doesn't fall into that type of recommendation, we get told to just make recommendations if DG was "necessary" and to propose some standards, but make sure there is a clear return on investment (ROI).

Usually this type of project for us is in an environment where the friction between this company's business and IT area goes back decades. The CIO absolutely feels he knows what the answer is and wants a rubber-stamped report. The designated sponsor can't stand IT or consultants and says we should spend the few weeks of the assessment "doing reports they need." Do you think DG is front-of-mind in this scenario?

Similarly, there is the request to assist after an artificial intelligence (AI) or analytics effort has failed to produce the expected miracles. (Most of them end up this way, by the way—ignore the hype.) As of the year I am writing this, even IBM has said that over 50% of its analytics efforts don't take hold due to data quality issues. But when we move toward DG, the same result. *We want the benefits, but not the work.*

Sometimes, especially in recent years, there is an improvement. Sometimes we are told "we know DG is very important, make sure it is included." But even the majority of those efforts hit a point where leadership says, "Wait a minute, we cannot make these changes—can't we just do AI or advanced analytics?" There develops a perception of overwhelming work or cost (not true).

The reason behind these scenarios is simply because of a lack of data literacy and will. The very thing most organizations have avoided is the reason they need expensive outside help to fix things. There is still a lot of education to be done.

DG is not part of a list of features. It is the underpinnings *of all of the possible solutions* to use data better.

Much of the work being done is a second or third pass at DG. Here is a recap of the many reasons DG programs get derailed.

- Data literacy—no understanding that is data is important, it is an intrinsic business capability.
- Business alignment and priorities—DG is a business capability. It needs to support business needs, and be prioritized like other capabilities that need to be introduced or evolve.
- Data quality—AI can be very scary, analytics will fall short, and internal risk goes too high when data quality is ignored. This doesn't stop organizations from disregarding data quality.
- Training and communications—The core components of successful change are training and good communications. Training usually is woefully inadequate, even for low profile efforts. Communications tend to be limited to some emails and then shouting!
- Stewards first—"We aren't getting any traction with management. We have to do SOMETHING. Let's appoint stewards." Not really. There isn't even a standard definition of steward in the industry, so what will they be doing in your organization?
- Technology first—I have beat this one to death. You should not buy anything until you have an operating model in place and have a realistic idea of pace and acceptance. Remember that 6-month period where things will not be stable?
- Classification as IT or applying IT thinking—"Data is IT's responsibility" is still a common refrain and still totally incorrect.
- Making it a project—I had a client whose CEO loved to talk about the great things to be done with data, and how the DG project was going to help the company. I could not get him to stop. So, one day I asked him how the "CEO project" was going.
- Hiring data governance—Luckily, I see less and less of this, but there is still a temptation for beleaguered executives to hire DG folks when they cannot get cooperation and resources from within. This is immediate termination of the program. In fact, when I see this, I recommend that DG is postponed.
- Change management—at the core, and last but not least, is organization change management (OCM). Truly, all of the other reasons can be dealt with through a good OCM effort. See Fig. 12.1.

FIG. 12.1

Managing change as the core of data governance challenges

Reminder that GOVERNANCE is not a new capability to business. Governance stems from the board of directors already. The only "new thing" to learn is that DATA (or information) requires governance. DG is, in a sense, *a refinement of what you already have.*

Let us revisit some of the critical items presented in this book. Each topic area we are revisiting is presented as a separate section to allow distribution or isolate a discussion.

Concepts

Information asset management (IAM)—Make sure you understand that "information as an asset" is not just a metaphor or brand. It means applying the same serious rigor that is applied to other "hard" assets. If you say you need DG, you are acknowledging belief in IAM. This gives you the right mindset.

The relationship of DG to information (or data) management—We introduced the concept of the "V" in this book. DG is *control*. It is the oversight and standardization component of enterprise information management. DG sets the rules and processes. Information (or data) management carries out the defined processes. It's important to keep the governing and managing of data separate. This leverages the concept of separation of duties. Your go-to analogy is that DG is to IM as accounting controls are to finance.

E for Enterprise—DG should never be considered as a project feature. There should never be two DG programs. Low profile efforts are planting the seed for enterprise capabilities. If not, you are just using DG to help do a project. DG is an enterprise program. The context of all DG discussions needs to be from an enterprise view. The roll-out is iterative, and each deployment is different. But the ultimate goal must be a level of enterprise-level adoption.

DG is a business program that delivers business capabilities—DG is *never* an IT program. It exists to provide the roles, rules, and controls for the data assets. It must be applied across the board to everyone in the organization. You are asking your organization to adopt or evolve business capabilities. Not IT capabilities.

The key aspect about understanding the DG flavor of capabilities is that there is a formal definition of roles for assigning responsibility and accountability for managing data and information assets. Just like all other assets.

DG capabilities are business capabilities—Rather than dive into processes and functions, any organization needs to fully know the WHATs. That is, what DG capabilities are required for the business to achieve its needs IF you use a capability approach to define and design your DG program, you will find that it fits a lot better into our organization, and developing an incremental roll-out plan is a lot easier.

Engagement model and operating frameworks—There is no such thing as a DG organization chart, at least in the physical, hierarchical sense. If you work in a hospital, do you have an organization chart for treating a patient? Not really, you apply the capabilities of many organizations.

The operating frameworks and engagement models are descriptions of work flows and communications that DG needs to operate. Base yours on some application of the "V" concept as well. No one can really have separate groups of staff doing DM and DG capabilities but try and get checks and balances. Remember DG has to define the "right things to do." Then DM does the right things. DG may likely identify those processes for "doing things right" as well—that is, the hands-on data management activities (the right side of the "V").

Information maturity—Information management maturity and similar benchmarks are your initial metrics. They make good measures of progress. They should never, ever be the business goals. The goals of DG are whatever DG needs to do to meet business needs. Also paying for an expensive assessment should be done as a means to fully understand what capabilities you have and where the gaps are. If you think you need an assessment to tell you where you are with DG, I can tell you with confidence—you aren't ready. Please do not do an assessment because management says "we should find out if we are not good at this data stuff for sure. Even having that conversation means it is a good bet you are immature."[1]

Evolution vs *revolution*—You need to learn how to govern. Humans are not born with these skills.[2] From the executive councils down to the operational activity, you must realize that behavior changes and education moves from the top to the bottom, and back up.

---

[1] I am perpetually amazed at organizations that get leaders together under the umbrella of a data-induced crisis. "Gosh, we better make sure we do not have a problem. Let's call the expensive consultants." Meanwhile, IT, AppDev, marketing, compliance, and data management are outside banging on the doors screaming "FIRE!"

[2] Actually, I think humans are wired to run as fast as they can from change, especially if the meeting invitation has the word "policy" in it somewhere.

Data governance means change—Top-to-bottom behavior changes require formal management via an OCM program. This is not really an option. Pursuing DG means you are not satisfied with what is going on. Something has to be different. Different equals change. So why not manage the change to the benefit of all the stakeholders? Formal OCM has been around for a long time. Allow OCM to work for your own DG effort. It is not intrusive. Low profile efforts tend to ignore change issues, then surprise sets in when even the most seemingly inoffensive change needs to be made and some middle manager has a fit.

The warning signs and useful tips

You are getting into deployment difficulties if you start to see these symptoms:

1. Quiet meetings—When right in the middle of all this new stuff, no one has any questions (you have some resistance brewing)
2. Philosophical soundings—Someone becomes a data expert out of nowhere and starts to question everything (this is more resistance)
3. Timing—Everything has been good until the last week, and now a flood of concerns and questions (guess what—this is also resistance)

There are resistance tips in Chapter 11 but to recap—deal with resistance as soon as you can. In the case of the three warning signs above, here are some tricks I have used.

1. Quiet meetings—say something absolutely nuts. For example, "Once we get the new glossary deployed, all paychecks will be withheld until all employees learn how to access the glossary portal." If everyone nods, demand everyone's attention and repeat what was said.[3]
2. Philosophical soundings—Take the expert aside, thank them for their intense interest, and ask them to review and comment on the training materials.
3. Timing—Bring the sponsor into a meeting to confirm that there will be no stopping the program. Reinforce that all questions are legitimate, and will be answered. Then once answered, we will proceed.

The value of data governance

At some point in time, the value of DG is either perceived as a traditional ROI (à la a project) or as a program that is required for the success of other programs or processes. This situation makes the statement of value for DG difficult. You can place the operation of DG into a model where a hard ROI is generated. But that is not a long-term number. The generation of a longer-term value requires the acceptance of DG as a program, and so the team must sell this to management.

With this in mind, the DG deployment team must create a business case and overcome these obstacles. If not, there will be no foundation for measuring success. A strong sponsor is important, but you must have the means to prove to the sponsor that it is working.

---

[3]True—once I had the top 5 executives in a really big company put their Blackberries in a shoe box on the table. That was risky and I got hammered for it, but it made the point.

The team needs to examine business opportunities and look for every opportunity to educate management on the importance of managing information as an asset. This will also address any long-term animosity between the business and IT.

Data monetization actually has become the main source of ROI, so we will touch that next.

Data monetization

At the start of the book we talked about data as an asset with value. We also pointed out opportunities that affect the bottom line through risk management, cost savings, and enabling people. Look up the six ways to monetize data (Fig. 8.3).

As of this writing a lot of work is being done in the context of data monetization. It all points to DG as the core capability to ensure monetization success.

Monetization of data is a surer means to justification vs traditional ROI. I am not a fan of ROI and I state that it should always be done. But ROI leans toward and implies project thinking. And DG, of course, is programmatic.

The critical success factors

For some reason, calling out three critical success factors are important to people. So here are the three most important success factors (out of many that we have covered).

1. DG needs to be set up to disappear—not vanish or stop, but to melt into the fabric of the organization. The DG "organization" is not a stand-alone department. Everyone must do governance once it is adopted.
2. If you do not manage the organization's behavior changes, you will not get DG to stick. DG requires OCM.
3. DG, even if started as a stand-alone concept, must be tied to an initiative. It is the best way to get visibility, try out policy, and designate targeted areas for training and orientation.

Some additional final thoughts

DG 2.0—As DG matures, there is a lot of talk about DG 2.0 What is really happening is maturity and addition of required capabilities. Personal opinion—there is no such thing as DG 2.0—I do not see any new, unheard of capabilities that were not needed before. There is new technology to support DG. But that means new customers for DG, not DG 2.0.

Data Debt—This edition introduced the concept of data debt as a metric of sorts for DG. Briefly, if you make decisions without considering the impact on, or use of, data (which is data literate behavior) there are costs. The costs occur in the future via dealing with lack of consistency, errors, redundancy, etc. Like any other future obligations, there is interest to be paid. It gets more expensive to fix later than now.

Using data debt as a guidepost is effective. A major financial institution used data debt as a means to fortify DG oversight of its application development areas. "We cannot afford more data debt from disjointed applications. A data glossary is to be maintained and followed to ensure consistency. *No development is permitted that uses files, tables, or data sources that are not in the glossary and no development can be done without maintaining the glossary."* The italics are the author's. This data debt-driven mandate worked, and the DG program got significant traction.

Infonomics—Doug Laney's aforementioned book has created significant discussion on how to look at data as a true asset in an accounting sense. Those discussions aside, DG is key to any formal recognition of data value.

Technology assets, people assets, data assets—Data and technology are separate assets, and need to be managed separately. They do not live in the same universe. Yes, "data" seems abstract. But it is common to say "people are our number one asset" in an organization. This is just as abstract and fully accepted.

The new economic component—For several thousand years Land and Labor drove economic activity. A few hundred years ago humans added Capital to that equation. We are now adding Data. We have similar types of ethical, administrative, securitization, and societal challenges and questions to be dealt with as when society had to understand the abstractions of capital markets, banks, and interest. We even have a new form of debt that can devastate an organization—data debt. Is data personal property? What are the ethics around data? How do we ensure privacy? Is surveillance capitalism an ethical economic model? DG must be present at every step as we work all of this out.

Elevator speech

Suppose you are on the elevator, but not with the CEO. We already talked about that. This elevator ride is with a reporter who wants to know if DG is worth a few minutes on a major network business episode. What do you say? Here are my thoughts.

- Companies want to embrace all of the benefits but not the work to get there. It is not overwhelming work, but it requires a few changes.
- Experience has shown the first change is to get the organization data literate. There are behaviors that people must change. But the goal is to have data management as much a part of organization language as budgets and risk.
- There are smart people in business, and this is not hard, but the literacy aspect means there is no focus on the data issues. There are guru platitudes, but literacy goes deeper.
- The key is Culture—the key to success with AI and analytics is to manage the people, then the data. Look at Costco—culture permeates the organization's success. Data requires the same.
- Data governance and management are market driven—you either have these capabilities in place or your business cannot achieve maximum potential.

Conclusion

Hopefully you can take something away from this edition. Remember, the first edition can be used as an adjunct. DON'T THROW IT AWAY.

There is plenty of data on things that have gone wrong due to lack of DG. A lot of experience has gone into this new edition.

The knowledge presented here must be applied all the time. Can we change the scenarios presented at the start of this chapter? I believe so. One thing about DG and DM—unlike other technology or business fads, they have not faded; for example, Where is your Business Process Reengineering expert? DG is still around because it is not a fad—it is essential, just suffering from some bad publicity and misunderstanding.

Organizations have to adopt principles and policies that address their longstanding abuse of operational data, and they will need to wean themselves off of inappropriate use of spreadsheets and Access databases. They need to understand that governance of data usage privacy and ethics are required to allow the 21st century opportunities that abound in data to flourish. AI will hurt people if there is no data quality. Advanced analytics will continue biased models without better oversight of algorithms.

DG must be more than just a bullet point on a slide for the leadership retreat. Will organizations have the will to make the change required? This remains to be seen. Hopefully, you have learned enough from this book to help your own organization be successful with DG.

Work areas, activities, and tasks

1

Data governance (DG) checklist			
Work area	**Topic**	**Activity**	**Tasks/considerations**
Engagement	Initiation	Obtain program approval	Obtain formal acknowledgment to start with a new program
			Communicate new program to leadership
		Develop DG rollout team structure, incl. stakeholders	Identify DG team and key stakeholders
			Identify DG steering body
			Perform SWOT analysis on participants
			Obtain team and steering body approvals and commitments
	Definition	Define DG and what DM is governed	Provide a straw person definition of DG to team
			Describe known areas of content for governance (if known)
		Identify business unit(s)—organizations subject to DG	List business units/divisions that may be subject to DG
			Identify key divisions in business units
			Understand significant strategies and initiatives
			Determine if divisional differences merit different DG
			Develop list of organizational units in scope of DG
		Identify capabilities that need DG (and don't have it now)	Identify obvious business capabilities that can benefit from data improvements
			Confirm addressing these capabilities with DG efforts
	Scope	Define scope and constraints with initial plan for DG	Define DG-specific tasks
			Define known constraints within proposed scope
			Define required assessments
			Define standard start-up tasks
		Approve scope and constraints	Review scope with proposed steering body
			Adjust based on feedback
			Develop final statement of DG scope

Continued

Data governance (DG) checklist			
Work area	**Topic**	**Activity**	**Tasks/considerations**
	Assessment	Information maturity	Determine scope of survey instrument
			Select or develop a maturity scale
			Identify all participants by name and group
			Orient respondents on importance and anonymity
			Agree on survey delivery (online, written, group focus)
			Review and modify maturity template
			Produce final form for delivery
			Deploy survey instrument
			Monitor online survey OR
			Distribute and monitor written version OR
			Prepare and deliver focus session(s)
			Collect and evaluate data
			Derive maturity score based on selected scale
			Collect existing standards, procedures, and policies for information management, info, resource utilization, prioritization, and controls, and map to IMM scale
		Change capacity	Determine the formality of the assessment; that is, an informal structured meeting format or a formal survey instrument
			Determine the target audience
			Define the survey population or interviewees
			Define the approach structured meeting, written or online
			Administer the survey or execute meetings
			Analyze and summarize findings
			Determine if additional investigation is required
			Leadership alignment
			Leadership commitment
			Determine what will be reported now vs sent to the DG team to use during subsequent phases
			Prepare change capacity report

Data governance (DG) checklist			
Work area	**Topic**	**Activity**	**Tasks/considerations**
		Data environment	Determine the formality of the assessment; that is, an informal structured meeting format, or a formal survey instrument
			Confirm data environment to be surveyed
			Obtain access and permissions
			Survey entire data environment—data supply chains, sources, movement, storage, disposition, metadata, staff, infrastructure
	Vision and plan	Identify obvious business benefits and metrics	Refine DG definition (if not defined elsewhere)
			List possible DG measures
		Describe new capabilities	Gather levers or stated goals and strategies and examine required content to enable them
			Identify obvious targets for improved quality or that would benefit from external scrutiny
			Examine significant business events and activities for content affecting risk such as safety, regulated products, rate filings, etc.
		Identify preliminary or obvious requirements	Gather existing artifacts, such as data or process models or DQ surveys
			Examine backlogs of report requests, web site updates, and requisitions for external data, data issues, anecdotal requests for DG
		Develop representations of future DG	Develop DG mission and value statement
			Present and refine mission and vision
			Obtain approval for mission and vision statement (if required)
			Build DG elevator speech
			Identify single page abstract of DG vision
			Identify notional DG touch points
			Develop "day-in-the-life" picture
		Complete DG start up plan	Adjust approach based on assessments and vision
			Define DG sprints or iterations
			Review and obtain approval
			Communicate final approach to stakeholders
			Kick off with formal event

Continued

Data governance (DG) checklist			
Work area	**Topic**	**Activity**	**Tasks/considerations**
Strategy	Alignment	Identify business needs	Gather/verify collected business goals and objectives
			Develop a list of known business initiatives, challenges, problems, and potential opportunities
			Turn challenges and opportunities into business directions
			Ensure each goal or objective is measurable
		Align DG with business needs	Gather/verify collected business goals and objectives
			Develop a list of known business challenges, problems, and potential opportunities
			Analyze measurable objectives for information and data requirements
			Identify how data can be used to support business objectives
			Determine data management and data governance capabilities that can support business initiatives
	Organization value	Determine core data principles	Use seed principles
			Apply GAIP™
			Align with existing enterprise principles and policies
			Add rationale and implications for each principle
			Select principles appropriate for initial use cases or opportunities
			Submit and approve principles to leadership
		Identify DG-enabled opportunities	Map DG opportunities to BIRs and key metrics to verify relevance
			Combine various opportunities into a source of value of DG
		Develop business value of DG	Connect data capabilities and issues with business needs
			Align DG opportunities with business benefits
			Identify potential cash flows from business goals
			Extract opportunities for using content and data
			Identify touch points where new managed content or data will touch, or be leveraged, to improve organization
			Isolate the processes that create value or achieve the goal related to the originating action

Data governance (DG) checklist			
Work area	**Topic**	**Activity**	**Tasks/considerations**
			Apply the various financial benefits and the costs to whatever benefit model is in use
			Create value statements of interaction of data and business goals
			Publish results to the DG team and/or steering committee
			Align business data needs with DG benefits (Show connection between business goal, required information, and data governance activity)
	Strategic requirements	Review existing business cases	Analyze existing initiatives, programs, goals, and strategies for strategic requirements
			Apply industry standard data requirements if necessary
			Verify that initial capabilities support strategic requirements
		Determine base line policy requirements	Draft initial policies from principles rationale
			Identify new policies that support new capabilities
		Confirm obvious capability areas	Present capabilities with related principles, polices, and values statements
			Match DG capabilities to existing business capabilities within business value areas
		Identify use cases to show value (as required)	Review alignment and strategic documents for specific opportunities where data must be used
			Verify that a real contribution to value will result
			Identify required DG and DM capabilities
			Document use case(s)
			Review and obtain prioritization of use cases
Architecture and design	Capabilities	Identify DG capabilities	Use standard list, prior lists of obvious capabilities, or develop your own list of WHAT needs to happen with DM and DG
		Align and prioritize capabilities with business needs	Match all capabilities with business capabilities that support strategy (or confirm what was done earlier)
		Identify tools and technology supportive of DM and DG capabilities	Cross reference DM and DG technology with capabilities and determine if new technology is required

Continued

Data governance (DG) checklist			
Work area	**Topic**	**Activity**	**Tasks/considerations**
	Operating framework(s)	Identify/refine DG processes	Identify DG processes
			Identify processes to sustain key business measures or metrics model
			Gather existing policies related to information or data management
			Identify processes to support standards, controls, and policy
			Identify processes to support master data and ERP projects
			Define issue resolution process
			Identify process to support AI, advanced analytics, and other algorithm-driven efforts
			Define/support regulatory drivers
			Identify any DG planning or management functions
			Identify requirements and processes for enterprise data model standards and procedures
			Identify requirements and processes for reference and code policies/procedures
			Identify processes to administer policies and standards
			Ensure processes and policies are not in conflict
			Optional: Work with Finance and Compliance and perform a pro-forma "Information Risk Forecast"
			Identify gaps in current state of data management
			Specify adequate controls
			Specify privacy and security concerns
			Specify compliance and regulatory concerns
			Identify changes to SDLC processes
			Design DG process details, deliverables, documentation for SDLC integration touch points
		Identify accountability and ownership	Examine processes requiring DG accountability
		Design DG operating framework	Identify business area touch points with DG functions
			Develop DG RACI from functional design
			Determine levels of federation

Data governance (DG) checklist			
Work area	**Topic**	**Activity**	**Tasks/considerations**
			Propose federated DG structure
			Identify layers of oversight based on RACI
			Determine how operations are organized
			Determine potential staffing
			Identify leadership of all levels
			Develop charters for main levels of DG operation
			Complete roles and responsibility identification at all levels
			Define or agree on role names (steward, custodian, etc.)
			Review and obtain approval of DG operating framework
		Design minimum sustainable operating model	Define initial DG operating layers
			Identify capabilities and processes of minimum impact and maximum value
			Reduce intended operating framework to only support the most essential capabilities
			Review and obtain approval of MSOM
	Engagement and workflow	Design required engagement models	Identify key internal processes to engage DG—planning, budgeting, SDLCs, compliance, etc.
			Define intersection and control points for DG
			Define work flow to support engaged processes
		Complete roles and responsibility identification	Define roles and responsibilities for specific DG roles (steward, custodian.)
			Develop accountability assignment approach
			Coordinate with HR and identified roles (e.g., data steward(s)) to revise performance goals and objectives
			Identify data governance oversight body(s)
			Council, forum, and committee members
			Identify specific contact points and protocol
		Socialize operating framework, engagement models, and workflow	Review and obtain approval of roles and accountability approach
			Develop data stewards identification template
			Identify data steward identification subject areas and prioritize them (e.g., customer)

Continued

Data governance (DG) checklist			
Work area	**Topic**	**Activity**	**Tasks/considerations**
			Identify roles (stewards, custodians, owners, etc.) by name
			Obtain approval of role assignments
			Initiate DG socialization
			Conduct orientation
			Review IM/DG principles with all roles
			Review operating models, MSOM, and engagement models
Implementation	Roadmap	Align DG with current efforts	Identify projects and stakeholders subject to standards and governance
			Gain support for DG engagement
		Assign DG to planned efforts	Engage product managers and project management
			Refine governance bodies and committees
			Refine DG charters
			Confirm stewardship and ownership model if necessary
		Define DG deployment increments	Develop DG management requirements
			Revise DG charter/mission if necessary
			Develop/refine DG positions
			Define roll out of DG to support data strategy or other initiatives and projects
			Define incremental roll-out of operating framework and engagement models
			Identify immediate governing tasks
		Develop short- and long-term deployment plans	Define DG roll-out tasks and schedule
			Define sprints and iterations
			Develop near-term and long-term views of roadmap
	Sustaining plan	Define sustaining requirements	Review/perform stakeholder analysis (or perform in parallel)
			Conduct an initial leadership alignment assessment
			Review other assessments
			Review or execute change capacity assessment (if not already done)

Data governance (DG) checklist			
Work area	**Topic**	**Activity**	**Tasks/considerations**
			Identify change management resources required
			Cross reference touch points, readiness, and stakeholder analysis
			Incorporate IMM results into the change capacity analysis
		Define organization behavior targets	Describe required engagement with policy
			Define nature and size of change
			Describe ability of sponsors to lead change
			Develop plan to engage sponsors (if required)
			Define timing of behaviors
		Develop change management plan	Determine level of detail
			Base plan on industry standard approach
			Define training requirements
			Define communications requirements
			Define conditions for sustainability success
			Define and design capture of sustaining metrics
			Identify change management teams
			Identify specific resistance profiles
			Develop responses to resistance
			Develop resistance management plan
			Review and approve resistance management plan
			Define staff reward structures/WIIFM
			Develop sustainability checklist
			Identify and design change measures
			Define feedback and monitoring approach
			Develop staff transition approach (use HR if necessary)
			Develop DG communications plan
			Develop DG training plan
			Prepare statement of change readiness

Continued

Data governance (DG) checklist			
Work area	**Topic**	**Activity**	**Tasks/considerations**
	Metrics	Define metrics for effectiveness	List metrics that show DG is demonstrating expected value
		Define metrics for efficiency	List metrics that show DG is improving efficiency
		Design metric collection and reporting	Identify sources for metrics
			Verify data can be collected, manually or from technology products
Operation and change	Deployment	Develop DG framework roll-out plan (of increments of operating framework)	Refine roadmap to identify smaller increments of the operating model
			Develop detailed project plan for roll-out
		Roll out DG framework	Complete new DG team identification/socialization
			Socialize DG program and area
			Socialize new DG managers
			Review DG charter(s)
			Present charters and DG principles to new staff and stakeholders
			Present sustaining activities and stakeholder analysis to DG staff
			Orient executive team to DG framework (if not done in sustaining activity)
			Schedule DG team, committees, and executives for their orientation, training, or educations
			Align DG team functions with road map projects
			Ensure estimates are understood and project management practices are in place
		Implement operational activity	Roll out initial DG functions
			Kick off initial stewards and projects
			Kick off DG operations
			Present initial road shows
			Publish guidelines and principles
			Implement DG policies/procedures orientation and training
			Publish and implement SDLC integration documentation

Data governance (DG) checklist			
Work area	**Topic**	**Activity**	**Tasks/considerations**
			Develop and conduct DG audit processes training
			Initiate DG audit processes
			Identify and define additional roll-out activity for the sustaining phase
		Implement operational support activity	Promote and interact with change management
			Perform and review audits and service levels
			Interact with governing bodies, data governance committees, and councils
			Perform operations and functions of DG framework—Data governance committees and councils
		Implement DG technology solutions	Verify supported capabilities and use cases
			Establish realistic and visible achievements
			Use best practices for tech installation
	Operation	Manage DG interactions with projects	Address problem areas aggressively
			Orient major project steering bodies
			Align DG project management activity with existing IT practices
			Identify project templates
			Identify DG project estimating tools
			Identify DG tracking and accounting procedures for IT
			Forecast DG project resources
			Utilize modified SDLC
			Ensure estimates are understood and project management practices in place
		Execute DG processes and activities	Align DG team functions with roadmap projects
		Complete and publish charters, policies, and standards	Integrate with current policy management process
			Utilize internal wiki, web sites, or other facilities to access artifacts
		Monitor attendance and engagement	Verify attendance of required individuals across all meetings and events
			Address falling attendance immediately

Continued

Data governance (DG) checklist			
Work area	**Topic**	**Activity**	**Tasks/considerations**
		Ensure technology is properly integrated into DG operation	Verify accuracy, integration, and performance of any technology solution
			Verify the operating procedures are followed, and training is effective
		Interact with data management activities	Alert data management area(s) of DG interaction
			Monitor participation and following of standards closely
	Measure	Implement metric collection and reporting processes	Verify you are following ordinary application and configuration policies if you are implementing measurement code
		Track distribution and use of metrics	Verify reporting and distribution is working and metrics are being reviewed
		Track use and issues around technology	Address technology-specific issues as soon as possible
	Sustaining activity	Verify leadership alignment	Perform leadership alignment checkpoint
		Assess organization impact	Transition staff to new roles (if required)
			Communicate short-term wins
			Communicate status and measurements of progress often to leadership
		Monitor sponsor effectiveness	Perform organizational impact analysis
		Monitor communication and training	Ensure communication plan execution
			Ensure training development and delivery
			Refine materials for training, orientation, road shows, etc.
		Transition DG to business	Develop additional advocates if necessary
			Manage implementation of DG checklist
		Manage resistance	Feedback and analysis of results
			Address any resistance as soon as discovered per the resistance plan

Change capacity assessment

The change capacity assessment identified can classify the types of issues the organization will have adapting to data governance.

Survey questions	Strongly disagree 1	Disagree 2	Neutral or undecided 3	Agree 4	Strongly agree 5
1	I understand the rational for and the focus of the upcoming changes					
2	Our senior leadership has communicated a clear and compelling reason for why change is critical to the organization's long-term success					
3	People in the organization feel they can speak candidly to anyone, even if their views are contrary to leadership's					
4	Previous changes in the organization have been well-managed (i.e., we have a good track record of managing change)					
5	I am confident that I will have the opportunity to express my opinions and make suggestions about upcoming changes					
6	I am confident that my opinions and suggestions will be given fair consideration					
7	I am confident that barriers to success of the change will be identified quickly and addressed					
8	Our senior leadership is aligned and committed to making the changes that will best position us for success					
9	The way people think and act in my work unit will be compatible with the changes that are determined to be necessary					

Continued

253

	Survey questions	Strongly disagree 1	Disagree 2	Neutral or undecided 3	Agree 4	Strongly agree 5
10	Although we haven't yet identified the specific changes that will be implemented, I trust that the organization will provide me with the resources needed to be successful in the future					
13	People throughout the organization understand the implications of the change for their areas of responsibility and feel that the change is <u>urgently needed</u>					
14	I believe that by improving our business processes and technology, I will be able to make more valuable contributions to the organization in the future than I can today					
15	I am confident that I will receive honest and accurate information about the change initiative and its impact					
16	People throughout the organization feel that changes to our business processes are urgently needed					
17	The risk and issues associated with the change have been identified in advance and appropriate actions have been taken to reduce their impact					
18	I am confident in my own ability to successfully implement any required changes that are identified					
19	I am the type of person that naturally embraces work-related changes					
20	I would be willing to function as a change advocate, helping my co-workers and business partners to embrace and implement the necessary changes					
21	I believe that the organization will be even more successful in the future as a result of the changes					
22	Our people have the skills, interest, and commitment to support the upcoming change initiatives					
23	Communication regarding the changes is open, direct, and regular					
24	Expected business results have been clearly identified up front, targets set, and measures established					

The information maturity assessment identifies how well the organization can manage and use information relative to externally defined stages. This is brief version targeted more to data governance.

		Strongly disagree	Disagree	Neutral or undecided	Agree	Strongly agree
		1	2	3	4	5
1	The enterprise has published principles on how we will view and handle data and information					
2	There are standards for how data is presented to all users					
3	There are policies for managing data that are published					
4	The data policies are understood and adhered to consistently					
5	There are rules for sharing and moving data in and out of the company					
6	There is a widespread understanding of the importance of data quality					
7	People are willing to be held accountable to a single standard of data quality					
8	Data controls are adequate enough so we trust all numbers and information that is published					
9	We can easily tie our need for information to specific business programs					
10	Our transaction systems have all the data we need to do reporting					
13	Data controls are adequate and we do not worry about regulatory issues due to data issues					
14	My department owns the data we use; that is, we are responsible for using it accurately and correctly					
15	The numbers my department reports to management are accurate					
16	The reports we produce sometimes disagree with similar reports produced in other areas					
17	The role of IT is to deliver data to me so I can analyze it and do all of my reports					
18	We use different people to do analytics than generate reports					
19	We have too many things to do to take time for data standards					
20	My company adapts quickly to changing business circumstances					
21	We are good at data analysis—very few decisions are "gut"					

Data governance charter template

Introduction

The charter is a critical document for a data governance program. It has several purposes:

1. Set out the operating framework
2. Document the purpose and objectives of the program
3. Identify the various components such as councils or sponsors
4. Establish the level of authority the DG operating bodies will have
5. Identify the type of federation
6. Identify the names of participants

Consider you may need a charter for each "layer" of your operating framework. A separate charter for sponsors, councils, and forums might be necessary in larger organizations.

The charter is a living document and should adapt to the growth and changes in data governance as the program evolves. Below is a basic outline featuring important aspects of charter documents, along with an explanation of each section.

Background

State what brought about the DG program. An MDM effort? Or a general program (EIM) to better manage all data and information?

Purpose of the DG charter

State the purpose of the charter. Is it emphasizing scope? Is there a DG Office that will oversee the program or is there an informal virtual operating framework? Is it describing all of the DG areas in one specific area? Describe the scope of the DG program.

Terminology

Very often a lot of new terms appear. Terms like MDM and data quality should be defined if they are key concepts integral to the DG program.

EIM vision and mission

If the DG program is a component of a general move toward enterprise information management then make sure the vision and mission of the EIM program is described. Describe the context of DG within the EIM program.

Objectives

Describe the *specific* measurable objectives of the DG program. What are the standards to be achieved the prove DG is working?

Reporting and metrics

Related to the objectives, what metrics are to be collected and reported?

Value proposition

Describe how the organization will be improved by implementing data governance.

IG operating framework summary

Describe the various arrangements and interaction of the organizational elements that will operate data governance. This means describing roles, responsibilities, and core processes.

Data governance council

One key area is the council that will essentially manage data governance. Describe the following key characteristics:

○ Touch points—Where will the council touch the organization?
○ Structure—Is it a formal hierarchy, a virtual body, a dedicated area (rarely)?
○ DG Council's Vision for Information Management—Describe how is the council supposed to view the formal management of information assets. This includes their:
  1. Roles
  2. Processes/Tasks
  3. Responsibilities
  4. Representation
  5. Subteams

DG office

There is usually a small coordinating body, usually virtual in nature, that acts as the permanent first point of contact for data governance. Even in the largest companies it is only a few people.

DG forums

Describe the forums, or the operating groups, that report to the council. This will be made up of stewards and custodians as well as personnel performing information management duties. Also describe their vision for information management.

1. Roles
2. Processes/Tasks

3. Responsibilities
4. Representation
5. Subteams

DG executive(s) or sponsor

There needs to be a sponsoring role. Describe this as you do the other roles, but make sure you also specify clearly the responsibility aspect before anything else. Sponsors have a tendency to fade away.

1. Responsibilities
2. Roles
3. Processes/Tasks

Logistics

This section describe how the data governance framework will executes its core operations.

1. Meetings
2. Voting
3. Communication

Authority

This section contains a clear statement as to the extent the data governance operating bodies can carry out the enforcement of standards. This section must be vetted to upper management and sponsorship and receive explicit approval.

Website

Describe the internal website(s) that contains information regarding DG, such as principles, policies, memberships mission visions, etc.

Document history

The charter is a living document. It needs to be flexible and easy to read. As such it will go through many changes.

Amendments

Data governance orientation and on-going knowledge transfer template

4

Three LEVELS of knowledge transfer for data governance		
1—Orientation	Understand vision, concepts, and value proposition so one can act and visibly is in support of change or activity	Master the WHY
2—Education	Ensure that the desired activity or change takes place from accountability and managerial view, what, why	Master the WHY and WHAT
3—Training	Ensure action takes place from view of those responsible for execution, "feet on the ground"	Master the WHY, WHAT, and HOW

Sample knowledge transfer or syllabus for data governance training			
Level	**Topic**	**Section name (useful to create reusable decks)**	**Description**
Orientation	**Managing data and information as an asset**	Concepts of data asset management	Concepts of data as an asset, terms and definitions
			General Vision of what IAM looks like
			Typical Mission, Vision, and Value statements
			Ramification and impact of data assets
			EDM solutions overview—MDM, BI, DQ, DG, AI, Analytics, etc.
Educations	**Data management program education**	Program overview	Concepts of operating enterprise data management (EDM)
			Operating vision for EDM
			EDM value proposition
Orientation	**Data Governance**	DG Concepts	Definitions, Value, and Concepts
			Value of DG
		DG Framework Requirements	Principles and Policies
			Best practices
			Intro to the operating framework for DG
			Intro data principles and policies

Continued

Sample knowledge transfer or syllabus for data governance training			
Level	**Topic**	**Section name (useful to create reusable decks)**	**Description**
Training	**Enterprise Data Governance and Oversight**	DG Orientation	DG Operating Framework
			DG Value and Vision
			Intro to the "V"
		Data Principles Orientation	Data Principles detail
	Organization Management	EDM & DG Operation Overview	Organization DG Framework (leadership, sponsors, councils, forums)
	Enterprise Data Governance and Oversight	Data Principles in Action	Data Principles in action
			Data Policies detail
		DG Operation	DG critical capabilities and Processes (Issues, Policy Change, etc.)
			Organization DG Metrics
			Moving up and down the "V"
			Organization DG Roadmap
	Sustaining Management	Organization Sustaining Requirements for EDM and DG	Organization EDM Change Management Overview
			EDM Culture Change Process
			EDM Maturity
			EDM EDG SWOT
			DG Risk Areas
			OCM Resistance approaches
On-going topics for clarification	**Ongoing orientation**	Concepts of data asset management	Concepts of data as an asset, terms and definitions
			General vision of what EDM looks like
			Typical Mission, Vision, and Value Statements
			Ramifications and impact of EDM
			EDM Solutions Overview—MDM, BI, DQ, DG
	Data governance value	Periodic updates on value-add of DG	Key metrics reporting
			DG brand and meaning
			Changes in business alignment
	Data governance compliance	Periodic review of enforcement and effectiveness of policy	Policy roll out
			Process effectiveness
			Issue resolution
	Organization change management	Progress of organizational change	Job changes
			Incentive progress
			Adjustments to incentives
			OCM metrics
			Ongoing surveys for OCM

Data governance capabilities and functions

5

This table replaces the function table in the first edition. This is not an absolute hierarchy. Capabilities can be customized or even combined for your organization. Therefore, the functions/processes can also be moved around. Most of the lines in this table have a blank cell. Feel free to insert your own process should your needs be unique.

Management activity type	Data governance capabilities	Data governance functions and processes			
Plan	*Data and governance strategy*				
	Enterprise information management (EIM) and data governance (DG) business alignment	Collect organization strategies and identify data opportunities	Understand business model	Identify DG capabilities to support organization	
	EIM and DG goal setting	Establish priorities for information projects	Understand goals for enterprise applications	Set DG goals—financial, maturity, organization	
	Compliance and privacy strategy	Establish role of DG with compliance	Set DG targets for compliance		
	Data ethics strategy	Identify ethics challenges	Facilitate data ethics discussions with leadership	Establish DG role in organization ethics	Create data ethics vision and strategy
	Data principles	Identify essential information principles	Confirm enterprise architecture (EA) principles with information principles	Apply data ethics to data principles and adjust	
	DG technology strategy	Coordinate with EA	Define the role of technology in DG	Assess current technology capabilities	
	Business strategy support	Consolidate cross-business unit business information needs	Socialize alignment and value of DG to leadership	Assist leadership in data monetization strategies	
	Data strategy support	Align DG capabilities with data strategy	Adjust DG roadmap to match data strategy	Support funding data systems	Review and approve DG aspect of data architecture and EA

	Applications strategy support	Align DG capabilities with applications development	Adjust DG roadmap to support applications strategy		
	EA support	Align DG capabilities with EA strategy	Adjust DG roadmap to support EA		
	Information management maturity (IMM)/ CMM strategy	Set targets for data maturity	Ensure maturity targets, business and data strategies are aligned		
Define	*Data governance requirements and design*				
	DG assessments	Assess information maturity	Assess current governance practices	Assess organization ability to change	
	Federation requirements	Identify common data needs across business units	Coordinate or consolidate cross organization needs	Segregate or classify content that may require federated treatment	
	DG scope and focus areas	Refine DG scope to match organization needs	Define areas of focus for DG if necessary		
	Capabilities definition	Define required capabilities	Prioritize capabilities as required		
	DG roadmap modification	Identify roadmap changes	Assess impact of roadmap changes	Approve and publicize roadmap changes	
	Controls specification	Examine regulatory capabilities for data control requirements	Examine financial capabilities for data control requirements	Specify data controls	Facilitate management review of controls

Continued

Management activity type	Data governance capabilities	Data governance functions and processes				
	Compliance identification	Examine compliance directives and define DG requirements	Identify DG capabilities and process for compliance	Define intellectual property policy		
	Enterprise risk management specifications	Facilitate risk-based requirements with risk management areas	Define required risk-related DG capabilities and processes			
	Ethics and privacy definition	Facilitate DG role in data ethics architecture	Define required ethics capabilities	Define data ethics framework (architecture and policy)		
	Policies and standards development	Extract policies from principles	Identify standards areas	Define enterprise data policy guidelines		
	Organization business information requirements (BIRs)	Define enterprise organization information requirements	Facilitate socialization and approval of enterprise BIRs	Define collaborative processes		
	Collaboration and communication setup	Define how the organization can best work cross functionally with data	Define principles for collaboration	Define collaborative mechanisms	Facilitate review of collaborative mechanisms	
	Governed metrics and measures	Define enterprise metrics and measures	Support creation of maintenance of metrics and measures	Facilitate review of enterprise metrics and measures		
	Metadata and model specification	Identify data models, analytic and AI models, and metadata DG requirements	Define standards for rules and models (data, analytics)	Define data meaning and business rules	Develop processes for management of models, algorithms, data, and metadata	Define enterprise metadata management environment

Taxonomy and ontology specification	Identify enterprise taxonomy and ontology requirements	Define taxonomy and ontology standards	Assist data management as required with taxonomy/ontology design	Define interface/interactions with ITIL if necessary	
Data lineage and provenance specification	Identify requirements for lineage and provenance	Define capabilities and policies to manage lineage and provenance	Integrate data quality needs into lineage		
Data classifications specification	Identify requirements for data classifications	Facilitate data classification definition	Define policy related to data classification	Define reference data requirements	
Data sharing specification	Identify data sharing requirements	Determine policy and technology constraints	Specify data sharing policy and process		
Data integration specification	Identify data integration requirements	Specify data integration policy and process	Define standards for ITIL integration		
Data life cycle management specification	Identify data life cycle requirements	Specify data life cycle policy and process	Define enterprise master data management (policies, design, processes)	Identify data supply chains	Define data quality approach—definition, profiling, and remediation
Data governance frameworks					
Operating framework definition	Define DG functions and processes	Define layers based on federation	Propose and adjust operating framework	Define change management plan to achieve operating framework	Create brand for DG program
Engagement model definition	Identify required engagement models	Define workflows and artifacts	Propose and adjust engagement models	Develop application development engagement requirements for reusability and consistency	Define change management plan to achieve engagement models

Continued

Management activity type	Data governance capabilities	Data governance functions and processes			
	Accountability and responsibility structure	Define responsibility and accountability for capabilities and functions	Identify roles and individual candidates	Refine DG processes after review	
	Data literacy	Specify data literacy needs	Define data literacy curriculum and goals	Define literacy targets	Design data literacy approach
	Roles and responsibilities	Supply details for new roles and responsibilities	Coordinate with human capital as required		
	Collaborative framework	Identify methods for cross unit data-related workflow	Define cooperative and collaborative processes	Establish communications mechanism(s)	
	DG technology requirements	Identify current technology to support DG	Define short- and long-term technology requirements	Establish direction for data management and DG technologies	Develop technology acquisition approach
	Supporting technology	Define additional requirements, architectures and standards for the following (if applicable) specific technology categories:			
	Metadata management				
	Data lineage/ provenance				
	Data mastering				
	Data security				
	Data life cycle management				
	Taxonomy ontology				
	Data movement and integration				

Reference data				
Data quality				
Data modeling				
DBMS				
Collaboration and knowledge management				
Stewardship				
Manage *Data governance operation*				
DG activity management	Create and maintain DG roadmap	Manage DG program	Mediate and resolve conflicts pertaining to data	Operate DG steering bodies, forums, and workflows
Data policy management	Approve new and refined principles and policies	Oversee adherence to policies	Review DG requirements with Compliance and Legal on a scheduled basis	
Data standards	Oversee standards utilization	Audit monitor privacy and security standards	Enforce standards for rules	Enforce standards for models
Metadata and glossary management	Develop and establish enterprise metadata management environment	Support maintenance of metadata and data models		
Measurement	Refine DG roll-out strategy and metrics	Measure and report progress of DG		
Issue management	Audit applications and other projects for DG compliance	Assess effectiveness of DG	Support issue reporting and resolution	Ensure issues are elevated properly and addressed

Continued

Management activity type	Data governance capabilities	Data governance functions and processes			
	Data access and user	Maintain policies for data use	Ensure business and organization access requirements are reflected in data requirements	Support or verify compliance with privacy and security policies	
	Content governance	Enforce enterprise MDM (policies, design, processes)	Enforce data principles, policies, and standards	Enforce document management standards for retention and life cycle	Apply retention policy as described
	Application development governance	Establish DG repository	Assist to establish priorities for IT projects	Validate App project data alignment with business and strategic needs	Track and leverage industry trends in EIM
	Compliance-related governance	Monitor and ensure data usage adheres to regulatory requirements	Facilitate compliance and data projects interactions		
	Security/privacy governance oversight	Review processes to support data privacy policy	Coordinate DG capabilities to support privacy and security		
	Ethics oversight	Audit data efforts and events subject to ethics policies	Monitor impact of data ethics policies		
	Leadership communication	Reach out to leadership and management in regular basis for feedback	Ensure communications events are taking place		
	Data risk oversight	Facilitate data risk management with internal risk managers	Review data risk impacts on regular basis	Report on data risk management	

	DG audits and controls	Facilitate reviews and control verification	Revise audit processes as required	
	Methods and workflow oversight	Oversee execution of engagement models	Identify potential issues from new engagement models	Solicit feedback on engagement model operations
	Policy administration	Coordinate with internal policy administration (if any)		
DG measurement				
	Effectiveness and efficiency metrics	Implement regular metrics and measurement of DG implementation	Review accuracy of DG metrics	Ensure metrics continue to meet measurement needs of DG
	Measurements of data quality and usability	Ensure data quality profiling and measurement continues	Support data quality community	Validate and facilitate leadership understanding of data quality results
	Business impact metrics	Call out business impact of data issues	Call out business benefits from data monetization	
	Data debt	Periodically survey for new data debt	Ensure all project and initiative communications address data debt	Support application development with data debt awareness and coaching
	Literacy and maturity targets	Review and adjust literacy and maturity targets	Report progress toward targets	Identify literacy obstacles and challenges to leadership — Facilitate impact analysis and resolution of literacy challenges
Communication				
Operate	Communicated expectations and accomplishments	Engage the communications plan consistently	Publicize achievements and recognize individual contributions	

Continued

Management activity type	Data governance capabilities	Data governance functions and processes			
	Communicated data-related directives	Ensure directives (standards and policies) are communicated correctly	Verify communication by seeing intended responses	Adjust communications vehicles as requires	
	Communication events and artifacts	Ensure events are held per the communications plan	Assist in event preparation	Prepare material for leadership participation in communication events	
	Training				
	Technology training	Deliver technology training	Monitor training accomplishments	Adjust training content and syllabus as required	
	Data literacy awareness training	Deliver data literacy training	Measure data literacy training effectiveness		
	Stakeholder and operations training	Deliver DG training according to training plan	Adjust mix of orientation, education, and training as required	Adjust training plan as required	
	Formal orientation and onboarding	Ensure new participants are onboarded to DG	Deliver consistent orientation and onboarding	Review and adjust onboarding as necessary	
	Data governance services				
	Data sharing agreement services	Assist stakeholders with internal data sharing agreements	Assist stakeholders with external data sharing, acquisition, and sale agreements		
	Data integration services	Define business requirements for information systems	Enforce information life cycle management (ILM) Policies	Enforce use of integrated and managed data	

	Data quality support	Define profiling rules	Refine remediation process and policy	Ensure DQ measures are distributed and considered	
	Data compliance and risk support	Review regulations for risk and compliance issues	Recommend DG capabilities for compliance	Assist with risk management planning	
	Data lineage and provenance support	Assist with lineage designs for applications or business processes	Support access to lineage or provenance metadata		
	Data ethics support	Review new initiatives for ethics challenges	Enforce intellectual property policy	Report annually to leadership on data ethics status	
	DG technology operation	Offer support for DG tools	Operate DG tools if required as a service	Identify situations where tools will aid projects and initiatives	
	DG technology delivery	Implement DG technology (only as a service—not as a developer)	Assist with technology testing and roll-out		
	Data delivery support	Review processes to support data access	Review processes to support data controls	Ensure data access is according to policy	Review data access to architectural standards to ensure compliance
Sustain	*Sustaining DG*				
	Organization change management requirements (OCM)	Define OCM requirements as required	Incorporate OCM needs into a sustaining plan	Determine extent and need for sustaining plan branding	
	Organization behavior changes	Identify specific changes in behavior with data	Create metrics or checklist to monitor behavior changes		

Continued

Management activity type	Data governance capabilities	Data governance functions and processes			
	Leadership alignment	Perform leadership alignment exercise periodically	Review alignment results with sponsors or leadership		
	Stakeholders management	Assess stakeholders periodically for engagement and understanding	Identify stakeholders who are changing support levels	Reassess stakeholders periodically	
	Conflict and resistance remediation	Facilitate understanding of resistance management plan	Identify conflicts and issues	Resolve issues and failure to execute correct engagement with DG	Provide efficient resolution of issues
	Change plan management	Keep change plan current and socialized	Adjust sustaining plan if OCM requirements change		
	OCM plan implementation	Roll-out sustaining	Monitor OCM activities for effectiveness		
	DG metrics	Ensure metrics show correct impression of DG operations	Publicize metric reporting to expand audience		
	Training oversight	Monitor training plan execution	Monitor effectiveness of training plan		
	Resource development	Develop new resources assigned to DG roles	Provide enhanced training to current resources as they expand responsibility		
	Promotion and branding	Monitor effectiveness of DG program brand	Refresh promotions and branding messages periodically	Assess effectiveness of DG program brand	
	Community and collaboration development	Assist in creation of cross-unit data communities	Support communities of practice for DG tools, policies, and capabilities	Monitor effectiveness of community and collaborative communications and interactions	

Stakeholder analysis

6

What is a stakeholder?	What is their role?	How will they react?	What will be their primary concerns?	What do we need from them?	How should we work with them?
A stakeholder is any organization or person that: ✓ Can influence the change ✓ Is affected by the change Stakeholders can be: ✓ Individuals ✓ Senior leaders ✓ Groups of employees such as IT or division managers ✓ Committees ✓ Customers ✓ Government or other regulatory agencies ✓ Brokers/ agents	Identify each stakeholder's role or roles. Will this stakeholder: ✓ Need to approve resources and/or decide whether the change can proceed (thus acting as a sponsor or gatekeeper)? ✓ Need to change as a result of the effort (a target)? ✓ Need to implement changes or convince others to change (an agent)? ✓ React to or judge the success of the effort? ✓ Need to be an advocate of the effort (a champion)? ✓ Perform work that can influence the success of the effort (a resource)?	How will the results of the effort be likely to impact the stakeholder? Will this stakeholder benefit or be adversely affected? Given the likely impact and prior behavior, how is this stakeholder likely to react? ✓ Vocal, visible support? ✓ Cooperative, quiet? ✓ On the fence? ✓ Say ok but be obstructive or complain behind the scenes? ✓ Express concerns vocally?	What are the primary concerns of this stakeholder? ✓ What do they need or expect from the change? ✓ What might influence whether they are supportive of the change? ✓ What will this stakeholder need to feel informed, involved, prepared, or validated during the change? ✓ What are the "red flags" or hot buttons for this stakeholder?	What do we need from this stakeholder? ✓ Approval/ resources ✓ Visible support/ public endorsement ✓ Access to them ✓ Access to people on their team ✓ Lack of interference with or blocking of the effort ✓ Information ✓ Task completion ✓ Flexibility ✓ Change in behavior	Given what we know, how should we work with this stakeholder? ✓ How will we prepare them for the change? ✓ How will we communicate with them? ✓ How will we address their needs/ concerns? ✓ Do we need to learn more about their needs, concerns, or likely reaction? ✓ Should they be part of the change team directly or indirectly involved (be a representative on the team, solicit input, or provide regular feedback)?

Leadership alignment assessment

7

Question	Purpose
What do you think the ultimate contribution of data governance (DG) will be for your organization?	Measures alignment around the DG vision, the organization's goals, and the true purpose of the anticipated changes.
What do you see as the major issues to successfully implementing DG? What can be done to address them?	Provides perspective on what's critical to successful implementation (specific actions, issues, or processes). Measures alignment around what needs to be done now to improve the chances of success.
Is DG an incremental or transformational change for the organization?	Measures alignment around perceptions of DG impact on the organization and the need for change leadership behavior from the leadership team.
What do you think are the best ways to encourage positive reception of DG by key stakeholder groups inside and outside of the organization?	Measures alignment around most effective approaches to stakeholder. Stakeholder groups could include branches, home office functions, service center, IT, producers, customers.
What is your definition of success for DG?	Determines alignment among the organization's leaders as to what success with DG means. Common definition of success drives common actions and behaviors.
Who's accountable for delivering on DG results?	Measures alignment around roles and responsibilities, and who is ultimately responsible.
What do you think your role is as a leader in making DG a success?	Measures alignment around leadership accountability and where authority for decisions will be situated.
What are your biggest concerns about the changes that DG will drive? How would you address those concerns?	Provides some insight into what the concerns are among the organization's leadership. Measures degree of alignment about areas of concern.
What other periods of significant change have you experienced in your time with the organization?	Gauges the change leadership experience/skill level of the organization's leadership overall. Provides insight into the organization's history with change and methods that have been effective or ineffective.

Communications plan

8

A Communications Plan should be developed early in the life of the project to ensure that communication needs are identified, and plans are established to meet those needs. The Communications Plan identifies who needs information, what information they need, the frequency and vehicles for communication, and the parties responsible for providing, consolidating, and disseminating the information. By providing a structured plan, we ensure that each stakeholder gets what he or she needs when they need it.

Event	Target audience	Purpose and objective	Timing frequency and location	Description and vehicles	Responsibilities— Sender, creator	Feedback mechanism
Provide the Name of the communication	*Detail the recipients of the information*	*Provide the purpose of the communication*	*Provide frequency of communication (e.g., one-time, every Friday at 10 a.m. in conference room a), date and if appropriate, a location*	*Describe the communication in terms of contents, format, delivery medium*	*List who is responsible for creating the communication as well as who is responsible for providing input*	*Describe the means to describe how the communications mechanism is working*
Executive Steering Committee Meetings	Executive Steering Committee	Update committee members on project status Approve EIM projects/initiatives Set direction for EIM and EIM	Monthly	Meetings, status reports	Executive Sponsor	Immediate discussion and comments captured in meeting minutes
Data Governance Council Meetings	Working Steering Committee	Update committee members on project status Resolve issues Confirm direction	Monthly	Meetings, status reports	Data Governance Council	Comments captured in meeting minutes

Continued

Event	Target audience	Purpose and objective	Timing frequency and location	Description and vehicles	Responsibilities— Sender, creator	Feedback mechanism
Data Management Committee Meetings	Committee Members	Allow the team to address issues relating to the quality of data and other data issues	As needed	Meetings	Committee Lead	Immediate discussion and comments captured in meeting minutes
		Provide direction and decision-making at the stall level				
		Forum for escalating issues to DGC				
Executive Cascade	EIM SC	Key points, action items, and to-do's for cascading key EIM Project messages down to their organizations	Monthly	Email or SP	Director, EIM	Event Feedback Form
	DGC					
Executive Toolkit	EIM SC	Slides, Scorecard, other files and to-do's for cascading EIM Project messages down their organization	Monthly	Email SP	Director, EIM	Event Feedback Form
	DGC					
EIM Project Team Meetings	Team Leads	Review Status	Weekly	Meetings or Conference call	Project Manager	Action items, decisions, and status as captured in meeting minutes
	Project Manager	Review Issues				
	Team Members	Identify, analyze, and mitigate project risks				
OCM Meetings	Sustaining Mgmt. team	Review status of sustaining activities	Monthly (after Comm. Plan, Education Plan are completed and accepted by Sponsor)	Meeting or conference call	OCM Lead	Action items, decisions, and status as captured in meeting minutes
	Project Manager	Identify issues and risks				
	Sponsor	Fine-tune plan based on progress				
"Did You Know"	All Stake-holders	Promote tidbits and new information about data quality and data governance	Weekly	Farfel web	EIM Team	Embed questions and opportunities to win prizes for those who visit the portal to review information
		The "hooks"—What's in it for me?				
Monthly DG Update	Data Governance Committee	Where is DG in terms of status, progress, and maturity?	Monthly	Metrics and status report	DG Team, Data Management Committee	Review instances for completeness

Event	Target audience	Purpose and objective	Timing frequency and location	Description and vehicles	Responsibilities— Sender, creator	Feedback mechanism
Monthly EIM Update	All stakeholders	What is complete and where EIM is with the transformation, maturity, and update of the DW Roadmap and DW projects	Monthly	Newsletter (create list of EIM staff and key stakeholders)	EIM Team	Review instances for completeness
Data Steward Forums	Data Stewards	Allow the team to discuss tips and techniques for managing data quality Obtain direct input from forum lead on issues and concerns Information sharing	Quarterly	Meetings	Manager DQ	Action items, decisions, and status as captured in meeting minutes
Leave behind	All employees	Keep IG/IM in front of mind.	As needed	Mouse pad Brochure Mug	OCM Team	Review instances for completeness
Public reminder	All employees	Keep IG/IM front of mind	As needed	Posters	DG Team, Data Management Committee	Review instances for completeness

Training plan

This is a very comprehensive example of a training plan. Yours can certainly be smaller as required. This is similar to the overall knowledge transfer plan in Appendix 4 - you will rarely need both.

Track	Track	Topic	Unit	Class #			Module name	Abstracts	Audience	Date
				Level	Unit	Part				
100	Data Management Fundamentals	Enterprise Information Asset Management (IAM)	n/a	100	001	1	Concepts of IAM	Concepts of IAM, General Vision, mission value prop. & ramifications of EDM, Definition of EDM solutions (MDM, etc.)	Custodians, Stewards, Councils	
200	Enterprise Data Governance (EDG) Fundamentals	Enterprise Data Governance	n/a	200	002	1	DG Concepts	Definitions, Value and Concepts	Custodians, Stewards, Councils	
				200	002	2	DG Framework Requirements for ACME	Principles and Policies; Best practices, Intro to ACME DG Framework	Custodians, Stewards	
300	ACME Enterprise Data Management (EDM) and EDG Knowledge Transfer	ACME DG Orientation	Basic Program Overview	300	101	1	ACME EDM Program Overview	Concepts of ACME EDM, ACME Vision, value prop. at EDM	Custodians, Stewards, Councils	
		Enterprise Data Governance and Oversight	Data Governance Processes, Organizations	300	102	1	DG Orientation	ACME DG Framework, incl. Principles, Value, and Vision, into to the "V"	Custodians, Stewards, Councils	
			EDM Guiding Principles, Supporting Policies				EDM Principles Orientation		Custodians, Stewards, DGC	
			Data Governance Processes, Organizations	300	102	2	DG Operation	ACME DG Road Map, Policies and Measurements	Custodians, Stewards, DGC	
			EDM Guiding Principles, Supporting Policies				EDM Principles in Action	Framework, Critical Process Review, the "V"	Custodians, Stewards, DGC	

Section	Topic				Title	Concepts / Content	Audience
Organization Management	Enterprise Information Architecture and Management Organization	300	116	1	ACME EDM/ACME EDG Organization Overview	Concepts, Roles Names of EDM and EDG Organizations	DGC, Stewards
Sustaining Management	Initial Organizational Change Management	300	117	1	ACME Sustaining Requirements for EDM/ACME EDM Overview	ACME EDM Change Management Overview, Process, Maturity, SWOT, Risk areas, Resistance management approaches	DGC
ACME EDM Road Map	ACME EDM Business Case/Alignment	300	125	1	ACME EDM Road Map Orientation	ACME Road Map overview, maturity levels, metrics, success criteria	DGC, Stewards
	ACME EDM Success Measures						
	Recommended Support/Application Projects				ACME EDM Projects	EDM project overview	DGC, Stewards
	Incremental Phased Roll-Out Timeline				ACME EDM Project Update		

Post-roll-out checklist

	Yes	To some degree	No
Are leaders acting as sponsors for and supportive of the new environment?			
Is the staff excited about the changes that are coming?			
Is there a safe outlet for feedback—reactions, concerns, and comments—for everyone?			
Is there adequate support for people to do their jobs effectively?			
Do people have time to do their jobs effectively?			
Does the organization have the skills/competencies to get the job done?			
Has the organization been trained in the new skills/competencies they require?			
Have competence and capabilities been built effectively so that objectives are met and results are achieved?			
Are there comparisons of progress against metrics and targets?			
Have new performance measurement and reward systems been implemented?			
Are we tracking performance that achieves results?			
Do we publically recognize individuals who demonstrate desired behaviors so that objectives are met and results are achieved?			
Do we acquire and place talent in a way that ensures objectives are met and results are achieved?			
Is the organization structure appropriate for the future state?			
Does our operating model ensure that objectives are met and results are achieved?			

Sponsor guidelines

Sponsor guidelines and expectations

As a sponsor, can you speak to all of these?

1. How do you define success for xxxxxxxxxxx? If successful, what will be different?
2. Why is the data governance program needed? How critical is it that we change? What are the business drivers for the change? What will happen if we don't change?
3. Why should our people want to change? What should we tell them, so they understand the importance of a data governance program for them and not just for xxxxxx?
4. What other changes have our people currently or recently experienced? Are they ready and capable of handling the data governance program effectively?
5. What will our leaders/managers need to do for the change to be successful during the roll-out of data governance and once it is operational?
6. What will our people need to do differently as a result of data governance? What specific behaviors will they need to exhibit?
7. What should you and the other sponsors of the data governance program do to make the changes successful? What should you be careful to avoid?
8. How successful has this organization been in implementing prior large-scale change? What worked? What didn't? What can we learn from this?
9. Who in the organization has been through large-scale change at another organization? What was their experience? What can we learn from them?
10. What are the potential obstacles to success? What do you think should be done about those?

Sponsor self-assessment

There are some basic areas a sponsor needs to master to be effective. Please score yourself (this individual) as a sponsor or business lead on the following statements. Use values of 1 through 5, with 1 being the lowest and 5 being the highest. A coaching plan can be developed based on the results. Scores between 40 and 50 equal strong or expert knowledge about leading change; scores between 30 and 40 equal midlevel knowledge of leading change; scores below 30 equal minimal knowledge of leading change.

Knowledge of change management processes and principles. _____

Understanding and support of the xxxxxxxxxxx program. _____

Able and willing to be an active and visible sponsor of the change. _____

Experience and success rate as a sponsor/business lead of past change projects. _____

Ability to communicate the vision and need for change to employees and managers.   _____

Degree to which the organization (employees and managers) would listen to and respect communications and support from this business leader.   _____

Ability to influence and build support with other business leaders.   _____

Ability to provide resources and funding for the project.   _____

Degree of direct control this sponsor/business lead has over the people and processes being impacted by the change.   _____

Degree of direct control this sponsor/business lead has over the systems, and tools being impacted by the change.   _____

Overall score_____

Leadership coaching

12

Below is an outline of a process and deliverable to assist leadership in being better data governance (DG) sponsors

Coaching plan objectives

- Knowledge transfer
  - Characteristics of DG-related changes
  - Specifics of this organization's program, including business value and priority
- Facilitate actions
  - Identify behavior and messaging to ensure on-going support
  - Identify activities to reinforce abilities of sponsors and leaders
  - Identify symptoms of success and resistance across various communities of stakeholders
  - Identify techniques to make recommendations to peers and stakeholders

Approach
Three sessions across three weeks
Deliverables

- Knowledge transfer material—A common orientation deck for all leaders being counseled
- Sponsorship recommendations—Recommended changes in sponsor activities to increase effectiveness
- New behavior and messaging—Agreed upon messages and behavior expectations to present to stakeholders
- Adjustments to roles and responsibilities—Recommended changes to program roles and responsibilities to increase program effectiveness
- Recommendation for new or adjusted messages
- Sponsors toolkit:
  - Guidelines to identify symptoms of success and resistance across various communities of stakeholders
  - Techniques to make recommendations to peers and stakeholders
  - New activities to reinforce abilities of sponsors and leaders
  - Enhanced education and orientation material to manage change

Resistance plan template

13

Adjust this for your own situation.

	Type of resistance				
	Active support/ advocacy	**Passive support**	**No resistance/ no support**	**Passive resistance**	**Active resistance**
Symptoms to look for	Leadership permits resource reallocation	Stakeholders show up but only offer support or participate when asked, or told to by a supervisor	Stakeholders (who are usually appointed) show up but do not contribute	When a stakeholder says they will get on board with the changes, but after they get something else done, they are possibly using prioritization to delay dealing with the change	Stakeholders do not even participate, and initiate all conversations with "that will slow us down"
	Stakeholders embrace new things and carry the message to their peers	Stakeholders embrace new things but only move out of comfort zone if others are doing so, OR if specifically asked by a supervisor		Resistance in the form of repeated questions, and repeating the same question, or seeming to not "get it" can be sign of resistance	Certain stakeholders have personalities where they vocally disregard new processes or standards when they feel they are unnecessary
				Stakeholders will often manifest uncertainty by foot dragging, or asking for more clarification than is reasonable	

Continued

	Type of resistance				
	Active support/ advocacy	**Passive support**	**No resistance/ no support**	**Passive resistance**	**Active resistance**
Mitigation	Learn what is important to each person and build support person by person				
	Move people along one notch at a time				
	Don't create unrealistic expectations—no change will resolve every problem				
	ALWAYS FOLLOW UP ON YOUR PROMISES TO GET ANSWERS—IF YOU DON'T, YOU WILL SHATTER TRUST AND CREDIBILITY				
	Do not make commitments that you cannot deliver				
	Pay attention to team members who have the strongest influence over other team members. Spending additional time with these individuals to move them up the continuum can have a major impact on the rest of the team			Assuming there is adequate education being offered, these individuals require clear goals and WIIFM to be willing to get through the period of uncertainty	You don't want to drive them underground into passive resisters. The short-term goal with any resister is not a sudden jump to supporter, but to help them accept the change— setting the stage for further buy-in
		Spend time directly asking for assistance and indicate how important they are. Also bring in supervisors if necessary		This situation requires leadership to reaffirm or shift priorities	Mitigating this situation requires leadership to reinforce the direction and emphasize any mandates that have been made
				Education needs to be adjusted to ensure that the necessary concepts have been conveyed and understood. This means measuring understanding; e.g., using quizzes and tests after training classes, as well as giving people time to make adjustments	

Guiding principles to policies alignment

14

#	Policies	Governance and accountability	Distinct communities	Right time, place, cost	Business alignment	Information asset quality	Risk management	Share and collaborate
1	Data, information, and content will be defined, classified and its usage tracked as though it were any other inventory item.			X				
2	Business alignment will be transparent to stakeholders across the information supply chains – IT and business area will understand the business reason and goals being achieved with data and content usage.				X			
3	Data and content will be assessed for risk. Potential risks will be identified and disclosed to appropriate parties.						X	
4	All associates are data stewards for the enterprise data assets. It is not acceptable to declare ANY data or content as "mine."	X			X			X
5	Formal information management will be a recognized business function with the required authority to ensure prudent, efficient, effective, and secure use of data resources.	X				X		
6	All activity that will use, develop applications, analyze, create, or alter data and content will use a consistent expression of meaning and rules, either via a model and/or taxonomy.	X						
7	There will be an active and effective data governance function.	X						
8	Various individuals will be accountable and responsible for data accuracy and quality. They will have the necessary authority to carry out the responsibilities.	X						
9	The data governance function will identify and approve the stakeholders, activities, processes, and policies for all data governance activity.	X						
10	Data governance will accommodate and support all defined enterprise architectures.		X					
11	Data governance will identify necessary data quality processes.					X		
12	Data governance will coordinate definition of data standards and policies, as well as define enforcement procedures.	X						
14	Projects and the PMO will ensure that data integrity (accuracy, quality, usefulness, and safety) are not compromised to achieve a project deadline.			X	X			
15	Data governance rules and enforcement will support development and sustaining of applications and other data and content solutions without excessive extension of schedules or budgets. Data governance will also enforce SOA / SaaS architecture standards.			X	X			
16	There will be a formal program to sustain and incorporate data governance into the culture.	X						
17	We will adopt and incorporate a global set of data principles that will guide the management of information and content as an asset.	X			X		X	
18	Data governance will work closely with corporate compliance and governance areas.						X	
19	Once a data source has been defined as "authoritative" or "certified" it will stand as the only approved "source of truth." This applies to sources of business events, report sources, and decision analysis or analytical data sources.		X					
20	We will maintain technology and processes to ensure consistent definitions, manage ease of navigation and administration, and track creation, use, and inventory of all data and content assets.			X	X			
21	Data governance will coordinate with compliance, regulatory, and risk management areas to ensure industry, state, federal, and international legislative and regulatory requirements will be adhered to.		X				X	

Data governance program start-up plan

15

While this example focuses on a quick strategy, you get the idea that you can boil down a lot of the work areas into just about anything you need to fit into your circumstances.

Data governance strategy and assessment plan

Resources	Data governance assess and direction setting — Work Area and activity	19-Mar	26-Mar	2-Apr	9-Apr
	Initiation				
	Project Team Confirmed				
BB	Team list and contact sheet	▓			
BB	Confirm meeting, interviews, and calendar		▓		
	Collected Documentation				
BB	Gather business strategies	▓			
BB	Gather data landscape info	▓			
	Final Project Plan and Approach				
JL	Agree on template for material and deliverable outlines	▓			
JL	On-site calendar		▓		
JL / BB / CC	Prepare kick-off	▓			
JL	Final project plan		▓		
	Strategy				
	Business / Data Governance Alignment				
JL / BB	Business interviews	▓			
JL / BB	Extract aligned DG capabilites from business plans	▓			
JL	Show alignment of DG capabilites to business strategy (the DG vision)			▓	
	Key Metrics and Strategic Data Needs				
JL	Identify KPIs and CDEs that can enable DG			▓	
JL	Identify other strategic data needs			▓	
JL	Identify tactical data needs			▓	
	Gaps in Current Programs to Future Needs				
BB	Define future DG capabilities			▓	
JL	Identify required DG artifacts to match capabilities			▓	
	Organization Capacity to Execute DG				
JL / BB	Determine DG execution capacity gaps			▓	
JL	Identify recommendations to close gaps			▓	
	Gap Analysis to Desired Level of Information Maturity				
JL	Define high level maturity target			▓	
JL	Identify broad steps to accelerate maturity			▓	
JL / BB	Final assessment of stakeholder's ability to move forward			▓	
	DG Value and Use Cases				
JL	State value areas for DG (confirm scorecard use case)			▓	
JL	Define/confirm use case for DG value demonstration			▓	
	Sustaining				
	Prepared DG orientation and education materials				
BB	Develop immediate orientation from templates				
JL	Identify high level education needs for DGC				
	Delivered DG Orientation and Training				
JL / BB	Deliver high level education to initial stakeholders				▓
JL / BB	Deliver orientation to leadership				▓
	Project Management				
JL	Develop client status		▓	▓	▓
JL	Develop internal status		▓	▓	▓

Sample data governance metrics

16

I look at the measurement of data governance success across two broad categories. Effectiveness, and Progress

Metrics to support data governance appear below in the BOLD outlines within the particular categories

Effectiveness - is the data governance program adding value in some way to the organization, i.e. how do we know it is effective? It is easy to see that initiatives like artificial intelligence (AI) or advanced analytics do better when there is data governance.

Enterprise value areas

Data governance supports organization improvements in a variety of ways. Use traditional business measures to show the difference between life with or without data governance.

Monetization of data - Data governance supports data monetization through analytics or AI.
- Change in key balance sheet items
- Change in key income statement items

Traditional initiatives - Measure impact on the balance sheet and income statement when data governance supports an initiative using MDM, data quality, BI, etc.
- Change in key balance sheet items
- Change in key income statement items

Organization improvement - The organization improves tangible, but non-financial statement areas
- Operating income per Knowledge Workers
- Inventory turns
- Supply chain cycle times
- Counts of key domains-customers, items, etc.
- Information maturity

Enterprise risk areas

Data governance can play a key role in risk management. Most organizations do some sort of risk assessment and estimate risk impacts, sort of like an insurance company. Your metrics observe changes in risk potential.

Operational risk - Data can prevent risks to everyday operations. Data governance supports consistency in metrics as well as enterprise risk management policies
- New business continuity via DG policies
- Employee turnover
- Breaches - privacy, security

Strategic - Measure the impact of data governance on managing strategic risk.
- Change in market share
- Improvement in technology capabilities
- Long term financial position / credit
- Improvements in reserves
- Change in reputation

Regulatory-Data governance is key in reducing regulatory risks, not only through penalties, but also fees, internal costs for discovery, and ongoing costs of maintaining compliance
- Regulatory Penalties
- Litigation fees
- Compliance costs
- DG / Compliance cost / total income
- DG / Compliance cost / risk reserves.

Efficiency , cost of ownership

Data governance can lower the cost of development and ownership of data systems. Thee metrics cover a variety of the elements of data cost of ownership

Overall technology costs - These are a series of ratios that indicate how efficient IT is at managing data
- Total cost of IT / Party (Customer, Member, etc.)
- End User Labor / Number Users
- Total BI/DW Budget / Total Users

Data handling - These are a series of metrics that help show how data governance can improve data movement and management
- Number of interfaces, external and internal
- Cost per interface
- Data errors (and all other data quality metrics)
- Data Budget / TB
- External data file costs

Benchmarks - Benchmark metrics are becoming more and more common as more organizations measure the impact and costs of data management and governance
- Number and types of tools
- Maintenance budgets
- License costs
- Training costs
- Industry maturity and other collected items

Progress - Is the data governance program moving ahead as planned? i.e. are we addressing challenges, issues, and helping stakeholders move forward?

People
- # decisions backed up by the leaders
- # of approved projects to oversee
- # of issues escalated and resolved
- # of data owners identified
- DG adoption rate by company personnel (Survey)
- Productivity increases

Process
- # of data consolidated processes
- # of implemented standards & processes
- # of consistent data definitions
- Dispute escalation timing
- SDLC integration performance
- Project schedules and costs

Technology
- # of terms mapped to data models and objects
- % completion of attributes
- % completion of glossary
- Improved reporting efficiency and accuracy
- # of data sources consolidated
- # of data targets using mastered data
- # of spreadsheets used
- Data integrity across systems
- Improved traceability of data
- Usage of a unique identifier

Phase activity task table from first edition

17

This table is from the first edition, and can still be used to help you establish your own checklist. However, while not intended, it seems linear or waterfall in style, so was revised in the new edition.

Phase	Activity	Tasks	Outputs/subtasks	Subtask outputs
Scope (initiation)	Identify business unit(s)—organizations subject to data governance (DG)	List business units/divisions that may be subject to DG	Business area candidates for DG	
		Identify key divisions in business units	Divisional candidates for DG	
		Understand significant strategies and initiatives	High-level business strategies driving DG	
		Determine if divisional differences merit different DG	Scope drivers of DG	
		Develop list of organizational units in scope of DG	DG program scope	
	Propose scope and initial plan to define and deploy DG	Define DG specific tasks	DG tasks	
		Define known constraints within proposed scope	Known constraints (e.g., market, time, regulations)	
		Define required assessments	Required assessment tasks	
		Define standard startup tasks	Standard enterprise program startup tasks (if any)	
	Develop DG roll-out team structure	Identify DG team and key stakeholders	DG team and stakeholder list	
		Identify DG steering body	DG steering body names	
		Perform SWOT analysis on participants	DG participant SWOT analysis	
		Obtain team and steering body approvals and commitments	Approved DG participants	
	Approve scope and constraints	Review scope with proposed steering body	Proposed DG scope	
		Adjust based on feedback	Feedback adjustments	
		Develop final statement of DG scope	Final DG scope statement	

Continued

Phase	Activity	Tasks	Outputs/subtasks	Subtask outputs
Assessment	Information maturity	Determine scope of survey instrument	Survey audience/areas	
		Select or develop a maturity scale	Survey maturity scale	
		Identify all participants by name and group	Survey participants	
		Orient respondents on importance and anonymity	Survey orientation	
		Agree on survey delivery (online, written, group focus)	Survey delivery method	
		Review and modify maturity template	Approved survey contents	
		Produce final form for delivery	Final survey	
		Deploy survey instrument	Survey available	
		Monitor online survey OR	Managed survey data collected	
		Distribute and monitor written version OR	Managed survey data collected	
		Prepare and deliver focus session(s)	Managed survey data collected	
		Collect and evaluate data	Survey database	
		Derive maturity score based on selected scale	Proposed information management maturity (IMM) score	
		Collect existing standards, procedures, and policies for information management, info, resource utilization, prioritization, and controls, and map to IMM scale	Mapped IMM to current state—gap analysis	
		Prepare findings for presentation	IMM survey presentation	
	Change capacity	Determine the formality of the assessment; i.e., is an informal structured meeting format, or a formal survey instrument	Change capacity survey format	
		Determine the target audience	Change capacity audience	

Phase	Activity	Tasks	Outputs/subtasks	Subtask outputs
		Define the survey population or interviewees	Survey audience/areas	
		Define the approach structured meeting, written or online	Survey approach	
		Administer the survey or execute meetings	Administered survey	
		Analyze and summarize findings	Survey database	
		Determine if additional investigation is required	List of business leaders requiring verification of support	
		Leadership alignment	Interviewed key individuals	
		Leadership commitment	Interviewed key individuals	
		Determine what will be reported now vs. sent to the DG team to use during subsequent phases	"Need to know" findings	
		Prepare Change Capacity report	Change Capacity report	
	Collaborative readiness	Determine the assessment's scope. Does it include:	Collaborative readiness assessment scope	
		Websites and content		
		Documents and sharing		
		Seeking and identifying existing communities of practice or interest		
		Workflow		
		Collaborative products		
		Contemporary facilities like instant messaging, texting, Twitter, or Facebook		
		Determine scope of survey instrument	Survey scope	
		Determine assessment approach interviews, document review, survey, or combination	Collaborative readiness assessment approach	

Continued

Phase	Activity	Tasks	Outputs/subtasks	Subtask outputs
		Collect existing standards, procedures, and policies for document sharing, workflow, internal wikis, blogs, etc. for review	Assessment source material	
		Collect inventory of SharePoint, notes, or other workshare facilities	Assessment source material	
		Identify all participants by name and group if necessary	Collaborative readiness survey participants	
		Orient respondents on importance and anonymity	Orientation for respondents	
		Identify interview of focus group participants if necessary	Focus group names	
		Produce final form for delivery	Final Collaborative survey instrument	
		Deploy survey instrument	Executed survey	
		Monitor online survey OR	Executed survey	
		Distribute and monitor written version OR	Executed survey	
		Prepare and deliver focus session(s)	Executed survey	
		Collect and evaluate data from surveys, documents, and meetings	Collaborative readiness survey database	
		Develop collaborative readiness statement based on predetermined scale	Collaborative readiness "score"	
		Prepare findings for presentation	Collaborative readiness report	
Vision	Define DG for your organization	Define information asset management (IAM) for enterprise (if not defined elsewhere)	Definition of DG/IAM philosophy—Draft brief impact and considerations document	
		List possible DG measures	Initial list of DG metrics	
		Develop DG mission and value statement	DG mission and value statement	
		Present and refine mission and vision statement	Refined mission and vision statement	

Phase	Activity	Tasks	Outputs/subtasks	Subtask outputs
		Obtain approval for mission and vision statement	Approved DG mission and vision statement	
		Develop straw person DG definition	Notional definition of DG	
		Build DG elevator speech	DG elevator speech	
	Draft preliminary DG requirements	Gather levers or stated goals and strategies and examine required content to enable them	Business goals affected by DG	
		Gather existing artifacts, such as data or process models or DQ surveys	Data artifacts affecting DG	
		Examine backlogs of report requests, website updates, and requisitions for external data, data issues, anecdotal requests for DG	Direct and indirect requests for DG	
		Identify obvious targets for improved quality or those that would benefit from external scrutiny	Data quality opportunities for DG	
		Examine significant business events and activities for content affecting risk such as safety, regulated products, rate filings, etc.	Risk areas benefitting from DG	
	Develop future representation of DG	Identify single page abstract of DG vision	DG vision statement	
		Identify notional DG touchpoints	DG business value proposition	
		Develop day-in-the-life picture	Day-in-the-life slide	
Alignment and business value	Leverage existing EIM (or other) business case	Review business documents, earlier findings	Business goals and objectives, findings from earlier activity	
		Confirm future relevance of goals and objectives to DG	Confirmed business goals relevant to DG	
		Confirm measures of goals and objectives	Metrics for confirming business goals	
		Clarify possible DG role in achieving business goals	DG roles in achieving business goals	
		Ensure each goal or objective is measurable	Confirmed metrics	

Continued

Phase	Activity	Tasks	Outputs/subtasks	Subtask outputs
	Align business needs and DG (if no source of business alignment)	Gather/verify collected business goals and objectives	Organization goals and objectives	
		Develop a list of known business challenges, problems, and potential opportunities	Categorized business goals, etc. into opportunities, challenges, problems	
		Turn challenges and opportunities into business directions	Business opportunities	
		Ensure each goal or objective is measureable	Confirmed objectives and business metrics	
		Convert levers, goals, and strategies to data requirements	Enterprise data requirements	
		Gather metrics, indicators, and other BIRs	Consolidated metrics and BIR list	
		Identify industry metrics (if not done yet)	Standard or industry metrics	
		Map DG opportunities to BIRs and metrics to verify model relevance	BIR/metrics to data model cross-reference	
		Optional—Map measures to source systems where DQ may be a concern	Metrics/BIRs to data quality issues cross-reference	
		Connect BIRs to data issues	Enterprise DG touchpoints	
		Build data usage/value worksheets if required	Usage value/info lever worksheets	
		Determine the business context to present benefits for DG	Enterprise value context	
		Schedule facilitated session with business leaders or subject matter experts	Business discovery session schedule	
		Capture business benefit results in the session, or refine results after presenting them	Discovery session results	
		Confirm future relevance of goals and objectives to DG	Confirmed business goals relevant to DG	
		Confirm measures of goals and objectives	Metrics for confirming business goals	
		Clarify possible DG role in achieving business goals	DG roles in achieving business goals	

Phase	Activity	Tasks	Outputs/subtasks	Subtask outputs
	Identify the business value of DG	Connect data issues with business needs	List of data issues cross-referenced with related business needs	
		Align DG opportunities with business benefits	DG opportunities to address issues affecting business needs	
		Identify potential cash flows from business goals	Business cashflow from affected business issues	
		Extract opportunities for using content and data		
		Identify touchpoints where new managed content or data will touched, or be leveraged, to improve business	Possible value points for new processes	
		Isolate the processes that create value or achieve the goal related to the originating action	Detailed actions in business processes achieving results through managed information	
		Apply the various financial benefits and the costs to whatever benefit model is in use	Financial benefit model for DG	
		Create value statements of interaction of data and business goals	DG value statement	
		Publish results to the DG team and/or steering committee	DG value presentation	
		Align business data needs with DG benefits (show connection between business goal, required information, and DG activity)	DG business value	
Functional design	Determine core information principles	Use seed principles	Initial list of information principles	
		Apply GAIP	Verification of principles to GAIP	
		Align with existing enterprise principles and policies	Adjusted and rationalized principles	
		Add rationale and implications for each principle	Draft enterprise information principles	
		Submit and approve principles to DG steering body	Approved information principles	

Continued

Phase	Activity	Tasks	Outputs/subtasks	Subtask outputs
	Determine baseline DG policies and processes to support business	Draft initial policies from principles rationale	Draft DG policies	
		Identify DG processes		
		Identify processes to sustain key business measures or metrics model	Metrics and BIR management processes	
		Gather existing policies related to information management	Existing IM policies	
		Identify processes to support standards, controls, and policies	Standards and controls for management processes	
		Identify processes to support master data and ERP projects	MDM and ERP DG processes	
		Define/support regulatory drivers	Regulatory DG processes	
		Identify any planning or management functions	DG planning and management processes	
		Identify requirements and processes for enterprise data model standards and procedures		
		Identify requirements and processes for reference and code policies/procedures	Reference and code DG processes	
		Identify processes to administer policies and standards	DG administration processes	
		Ensure processes and policies are not in conflict	Policy/process cross-reference	
		Optional: Work with Finance and Compliance and perform a pro-forma "Information Risk Forecast"	Information Risk Forecast	
		Identify gaps in current state of data management	Processes to close current DG deficiencies	
		Specify adequate controls	Data controls	
		Specify privacy and security concerns	Privacy/security controls	

Phase	Activity	Tasks	Outputs/subtasks	Subtask outputs
		Specify compliance and regulatory concerns	Compliance and regulatory DG processes	
		Specify key DG process flows	Define issue resolution process	DG issue resolution flow
			Define process for DG policy and standards changes	Policy and standards maintenance flow
			Define DG and project interaction	Project DG flow
			Develop new organization performance objectives	DG performance objectives for business areas
		Identify other DG detail processes	Identify changes to SDLC processes	SDLC change requirements
			Design DG process details, deliverables, and documentation for SDLC integration touchpoints	SDLC changes
			Develop revised process/ policy alignment plan (review/update existing policies and processes related to data governance and EIM)	Revised polices affected by governance
≈	Identify/refine IM functions and processes	Specify/identify IM processes	Revised IM processes (not DG)	
		Separate IM functionality from DG	Separate lists of IM and DG functionality	
	Identify preliminary accountability and ownership model	Examine processes requiring DG accountability	Accountability processes	
		Identify business area touch points with DG functions	DG touchpoints	
		Define preliminary DG operating layers	Preliminary view of DG operating layers	
	Present EIM DG functional model to business leadership	Prepare DG functional presentation	DG functional presentation	
		Gain acceptance of DG processes in principle	Approved function list	

Continued

Phase	Activity	Tasks	Outputs/subtasks	Subtask outputs
Governing framework design	Design DG operating framework	Develop DG RACI from functional design	DG RACI	
		Determine levels of federation	DG federation layers	
		Propose federated DG structure	DG federation model	
		Identify layers of oversight based on RACI	Organization layers for DG	
		Determine organization model	DG framework organization chart	
		Determine potential staffing	DG organization staffing	
		Identify leadership of all levels	DG leadership	
		Develop charters for main levels of DG organization	DG charters	
	Complete roles and responsibility identification	Define data stewards' roles and responsibilities	Stewards/owner roles and responsibilities	
		Develop data steward/ accountability identification approach	Accountability definition for DG	
		Coordinate with HR and identified data steward(s) to revise data steward(s) performance goals and objectives	Revised performance objectives for stewards	
		Identify DG oversight body(s)	DG oversight framework	
		Council, forum, and committee members		
		Identify specific contact points and protocol		
	Review and obtain approval of DG organization design	Review and obtain approval of data stewards identification approach with leadership	Approval to acquire stewards	
		Develop data stewards identification template	Steward template	

Phase	Activity	Tasks	Outputs/subtasks	Subtask outputs
		Identify data steward identification subject areas and prioritize them (e.g., Customer)	Steward content oversight areas	
		Identify stewards and owners	List of stewards and owners	
		Obtain approval of stewards and owners	Approved stewards and owners	
	Initiate DG socialization	Conduct data stewards' orientation	Completed orientation	
		Review IM/DG principles with councils and stewards	Principle review session	
Road map	Integrate DG with other efforts	Identify projects and stakeholders subject to standards and governance	List of projects and stakeholders subject to DG	
		Refine governance bodies and committees (if part of EIM)	Enhanced EIM oversight	
		Refine DG charters (if part of EIM)	Adjusted EIM/DG charters	
		Confirm stewardship and ownership model if necessary	Reviewed DG roll-out with stewards and owners	
		Define roll-out of DG to support EIM Road Map or other identified projects	EIM/DG Roll-out Road Map	
		Define DG roll-out tasks and schedule	DG roll-out schedule	
	Define sustaining requirements (only if not part of EIM)	Review/perform stakeholder analysis (or perform in parallel)	Stakeholder impact on sustainability	
		Review other IM assessments	Change readiness report	
		Execute change capacity assessment if not already done	Change readiness report	
		Identify change management resources required	List of resources for change management (team, facilities, tools)	

Continued

Phase	Activity	Tasks	Outputs/subtasks	Subtask outputs
		Cross-reference touchpoints, readiness, and stakeholder analysis	Change management areas	
		Incorporate IMM results into the change capacity analysis	Maturity/change capacity targets	
		Perform stakeholder analysis (if necessary)	Identify DG stakeholders	DG stakeholders list
			Perform SWOT analysis (all stakeholders)	SWOT by DG stakeholder
			Complete stakeholder analysis	Summarized conclusions from SWOT
			Review with DG leadership	Reviewed SWOT findings
			Determine levels of commitment for key stakeholders	Classified stakeholders
			Review results of stakeholder analysis with leadership (DG steering or sponsors)	Reviewed SWOT findings
			Determine action plan to address improving levels of stakeholder commitment	SWOT action plan (can be part of sustaining requirements)
		Conduct an initial leadership alignment assessment		
		Define nature and size of change	Sustainability scope and impact	
		Describe ability of sponsors to lead change	Sponsor ability report	
		Develop plan to engage sponsors (if required)	Sponsorship sustainability approach	
		Define training requirements	DG sustaining training requirements	
		Define communications requirements	DG communications requirements	
		Prepare statement of change readiness	Change readiness presentation	
		Complete requirements to sustain DG	DG change management and sustaining requirements	

Phase	Activity	Tasks	Outputs/subtasks	Subtask outputs
	Develop the Change Management Plan	Define conditions for sustainability success	Sustainable DG criteria	
		Define and design capture of sustaining metrics	DG sustaining metrics	
		Identify OCM team members	DG Change Teams	
		Identify specific types of resistance	DG Resistance Profile	
		Develop responses to resistance	DG Resistance Responses	
		Develop resistance management plan	DG Resistance Remediation Plan	
		Review and approve resistance management plan	Approved response to DG resistance	
		Define and align staff performance goals and reward structures	WIIFM statements	
		Develop sustainability checklist	Sustainability checklist	
		Identify and design change measures	DG change management success metrics	
		Develop staff transition approach (use HR if necessary)	Staff transition approach	
		Develop Communication and Training Plan (see below)	Communication and Training Plans	
		Develop DG Communications Plan	Identify audiences	DG communications audiences
			Create messages and branding	DG messages, branding
			Identify vehicles for communications	DG communications delivery mediums
			Define timing, frequencies, and delivery means	DG communications schedule
			Review and approval of Communications Plan	Approved DG Communications Plan

Continued

Phase	Activity	Tasks	Outputs/subtasks	Subtask outputs
		Develop DG Training Plan	Identify audiences	DG training audiences
			Identify levels and extent of training: orient, educate, train	DG training syllabus
			Identify vehicles for training	DG training delivery methods
			Define timing, frequencies, and delivery means	DG training schedule
			Review and approval of Training Plan	Approved DG Training Plan
	Define DG operational roll-out	Develop DG management requirements	Day-to-day DG management	
		Revise DG charter/mission if necessary	Revised DG charters	
		Develop/refine DG organization positions	Revised DG organization	
		Identify immediate governing tasks	Near term governance activity	
		Define DG roll-out schedule/road map	Road Map	
Roll-out and sustain	DG operating roll-out	Complete New DG team identification/socialization	Verified DG team socialized	
		Socialize DG program and area	Understanding of the DG team role to constituents	
		Socialize new DG managers	An operational, effective DG organization	
		Review DG charter(s)	DG charter	
		Present charters and DG principles to new staff and stakeholders	Oriented staff	
		Present sustaining activities and stakeholder analysis to DG staff	Oriented staff	
		Orient executive team to DG organization (if not done in Sustaining Activity)	Oriented executive team	
		Schedule DG team, committees, and executives for their orientation, training, or education	Training and orientation Relocation Job descriptions	

Phase	Activity	Tasks	Outputs/subtasks	Subtask outputs
		Align DG team functions with Road Map projects	DG managing projects	
		Ensure estimates are understood and project management practices in place	DG managing projects	
		Roll out initial DG functions	Kick off initial stewards and projects	DG program kickoff
			Kick off DG organization	DG program kickoff
			Present initial road shows	DG road shows
			Publish guidelines and principles	DG principles and policies
			Implement DG policies/procedures orientation and training	DG training
			Publish and implement SDLC integration documentation	SDLC changes
			Develop and conduct DG audit processes training	DG audit processes training
			Initiate DG audit processes	DG audits processes in place
			Identify and define additional roll-out activity for the sustaining phase	Additional activity as required
		Implement DG program metrics		DG metrics definitions
				DG/sustaining metrics comparison
				Metrics presentation
				Metrics collection mechanism
				A set of metrics that are deployed and being used to report on effectiveness of DG/IAM

Continued

Phase	Activity	Tasks	Outputs/subtasks	Subtask outputs
		Implement tools and technology		DG tools
		DG operations	Promote and interact with change management	Operational DG
			Perform and review audits and service levels	Operational DG
			Interact with governing bodies, DG committees, and councils	Operational DG
			Perform operations and functions of DG framework—DG committees and councils	Operational DG
	Execute the Change Management Plan	Communication plan execution	Communication events	
		Training development and delivery	Training events	
		Transition staff to new roles (if required)	Transitioned staff	
		Feedback and analysis of results	DG sustaining feedback	
		Perform leadership alignment checkpoint	Leadership alignment update	
		Perform organizational impact analysis	Impact of DG on organization	
		Manage resistance	Remediated resistance	
		Manage implementation of DG checklist	DG checklist	
		Refine materials for training, orientation, road shows, etc.	Refined materials	
		Develop additional advocates if necessary	Revised sponsors	
		Communicate short-term wins	Short-term win communications	
		Communicate status and measurements of progress, often to leadership	DG progress scorecard	
		Address problem areas aggressively	Issues resolution log	

Phase	Activity	Tasks	Outputs/subtasks	Subtask outputs
	DG Project Management	Orient major project steering bodies	Awareness of ongoing operation of DG	
		Align DG project management activity with existing IT practices	DG/IT practice aligned	
		Identify project templates	DG project templates	
		Identify DG project estimating tools	DG estimating	
		Identify DG tracking and accounting procedures for IT	DG tracking	
		Forecast DG project resources	DG resources	
		Utilize modified SDLC	DG enhanced work products	
		Interact with enterprise PMO (if one exists)	PMO DG interaction	
	Confirm operation and effectiveness of DG operations	Evaluate organization structure	Verified DG organization	
		Confirm effectiveness of jobs/people	Verified role transitions	
		Verify policies/procedures	Verified policies	
		Review incentives	Verified incentives	
		Monitor and report sustaining metrics	DG scorecard	
		Execute measurement surveys (if designed)	DG surveys	
		Hold focus groups/ interviews for feedback	DG focus group feedback	
		Execute change integration checklist	DG sustaining checklist	
		Change integration/ adoption assessment	Change adoption assessment	
		Realign impacted policies/ practices and procedures	Realigned DG policies	
		Revise staff performance objectives and reward structures	Revised incentives for DG	

Executive level sample roadmap

18

This final roadmap sample (others are in the text) would be suitable for a high-level overview.

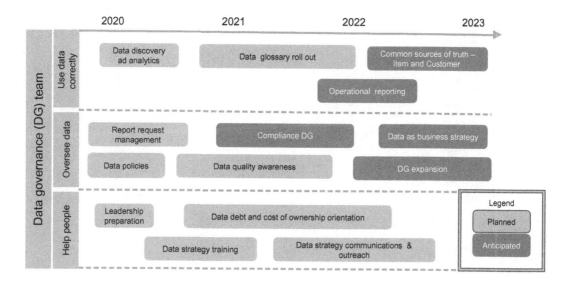

Index

Note: Page numbers followed by *f* indicate figures, *t* indicate tables, *b* indicate boxes, and *np* indicate footnotes.

Printed in the United States
By Bookmasters